Numerical Control and Computer-Aided Manufacturing

Board of Advisors, Engineering

Numerical Control and Computer-Aided Manufacturing

Roger S. Pressman
University of Bridgeport

John E. Williams
University of Connecticut

John Wiley & Sons
New York · Santa Barbara · London · Sydney · Toronto

Standards NAS-943 and NAS-955 copyright the
Aerospace Industries Association. Portions
reprinted by permission.

EIA standards RS-227A, RS-244A, RS-267A, RS-273A
and RS-274B copyright Electronic Industries
Association. Portions reprinted by permission.

Library of Congress Cataloguing in Publication Data:

Pressman, Roger S.
   Numerical control and computer-aided manufacturing.

   Includes bibliographical references and index.
   1.   Machine-tools—Numerical control.   2.   Manufac-
turing processes—Data processing.   I.   Williams, John
Ernest, joint author.   II.   Title.
TJ1189.P728   621.9'02   76-23218
ISBN 0-471-01555-5

Printed in the United States of America

10 9 8 7 6 5 4 3 2 1

To

Barbara P., Mathew, and Michael
Barbara W., Lorraine, and Stephen

Preface

The manufacturing process has undergone significant change during the past 20 years. The major contribution to this change has been programmable automation, made possible by machines that can be automatically controlled to perform a variety of tasks.

From humble beginnings in the late 1940s, numerical control has progressed to the forefront of automated machine development. In little more than two decades, numerical control has evolved from simple automatic positioning machines, controlled by instructions on perforated tape, to sophisticated machine systems integrated within a computer controlled manufacturing network.

An understanding of the numerical control process requires information from many engineering disciplines. This book presents a treatment of numerical control that bridges the gap between the hardware and software elements that enable the system to function. Numerical control is traced from its foundation in control theory, through the development of the hardware components of the system, into the use of computer based elements of modern programmable, automated devices.

The initial chapters of this book present a conceptual view of a conventional numerical control device. The fundamental concepts of automatic control theory are introduced to familiarize the student with basic mathematical methods and terminology. The elements of a numerical control system are next presented as interrelated components of an overall information network. The machine control unit and system elements that perform digital control, actuation, and monitoring are considered. Finally, special considerations for the design of numerically controlled machines are discussed.

In subsequent chapters the emphasis shifts from hardware to the interface between the machine control system and the coded data that drives the machine. A discussion of numerical control input and output concentrates on the nature of the data that provides the man-machine interface.

The methods that are used to generate coded instructions for input to a programmable controller vary greatly in their level of complexity. Simple manual procedures and sophisticated computer techniques are presented to illustrate the various modes of part programming. Special mathematical techniques, fostered by the expanding application of numerical control for the manufacture of complex components, are examined in a separate chapter.

The concluding chapters deal with numerical control optimization methods and computer-aided manufacturing. Adaptive control provides a second information path that optimizes numerical control operation by evaluation and modification of process parameters. Computer managed numerical control systems form the foundation for the latest step in the evolution of the manufacturing process—the computer based manufacturing system.

This book has been developed as a teaching tool for a subject that touches on many diverse engineering disciplines. The chapters have been arranged so that the physical elements of the numerical control system are explained before software and advanced systems are considered. Example problems are used to illustrate important concepts, and references are listed at the end of each Chapter to guide further investigation of various topics. Chapters Two to Nine contain problems for the student to solve.

Numerical Control and Computer-Aided Manufacturing is a suitable text for a one-semester course in mechanical, industrial, or manufacturing engineering at the undergraduate level. It may also be used to complement existing material in courses dealing with manufacturing processes developed for engineering technology programs. It provides a comprehensive framework for a graduate level course in programmable automation.

We acknowledge the many authors who have contributed to the numerical control literature during the past two decades. Their work and the publications, specifications, and manuals provided by numerical control manufacturers have had an important influence on the subject matter and methods of our presentation. We also thank the many contributors in industry who have provided state-of-the-art information and photographs.

Finally, we thank Barbara Williams for typing and Barbara Pressman for checking the manuscript. Both are thanked for their patience, understanding, and encouragement.

ROGER S. PRESSMAN
JOHN E. WILLIAMS

Bridgeport, Connecticut

Contents

Numerical Control and Computer-Aided Manufacturing

Chapter One
Introduction to Numerical Control

Man has been described as a tool using animal. Among the characteristics that distinguish him from other species is an ability to fashion complex devices that magnify or extend his own capabilities. These devices which we call machines have governed the rate of man's material progress throughout history. The evolution of the machine can be attributed to its inherent *propagating power.* Existing tools make possible the manufacture of more advanced tools which in turn serve to accelerate the evolutionary process.

1.1 Man and the Machine

The first machine tools are believed to have been developed more than 2500 years ago. These early rotary devices allowed the artisan to produce intricate circular forms from wood and other hard materials. Although the early machines extended man's ability to produce relatively complex shapes, it was not until the fourteenth century that the first elementary precision machines were developed. The mechanical weight driven clock, proposed by Giovanni DeDondi (1318–1389), became the impetus for the development of the first true machine tools, such as the screw cutting lathe. The advent of the industrial revolution greatly accelerated the evolution of the machine tool, and the development of the steam engine by James Watt in the latter half of the eighteenth century precipitated requirements for new devices and precision in metal cutting tools.

   In 1798 Eli Whitney signed a contract with the U.S. Government to produce 12,000 muskets and promised that the parts of each musket would be interchangeable. The commitment required manufacturing control which had never before been attempted. Whitney and his associates designed water powered machinery to perform the forging, boring, grinding, polishing, and rolling operations at his mill in New Haven, Connecticut. Although Whitney had contracted to produce the muskets in two years, only 500 were completed in September 1801, and it took him eight years to satisfy the contract. Nevertheless, Whitney had developed precision manufacturing methods which were a hallmark for his time.

The principles of manufacture first applied by Eli Whitney were implemented and enhanced by Samuel Colt and Elisha Root at their armory in Hartford, Connecticut. By 1855 no less than 1400 machines were installed. Cutting tools, jigs, and work-holding fixtures were specially developed by Root, and the Colt Armory produced weapons with the highest degree of precision attained at that time. As a historical footnote, two of Colt's employees, Francis Pratt and Amos Whitney, went on to found a firm that has become a major producer of numerically controlled machine tools.

During the latter half of the nineteenth century, important refinements were made to metal cutting machine tools. Early machines, such as the Lincoln Miller, illustrated in Figure 1.1, were continually updated to improve production rates. The desire for increased output led in 1873 to the development of the first *automatically controlled* machine, C. M. Spencer's automatic lathe. Using what he termed a "brain wheel," Spencer fostered the birth of automation.

By 1900 the American machine shop contained basic machine tools that were not very different in function and form from those in use today. The manufacture of automobiles in England and then in the United States served as a driving force for the development of better machine tools. By 1914 the Ford Motor Company was producing over one-million "Model-Ts" *each* year. In little more than fifty years mass production had become a way of life.

1.2 Numerical Control

This century has seen machinery become more automated, thereby eliminating machine-operator intervention in the manufacturing process. Yet, as automation has increased, machines have become inherently more specialized. A highly automated machine that may turn out 20,000 components per day, will generally be able to produce only a limited class of components. Until recently, prototypes and low volume components were produced by manually operated conventional machine tools.

With the advent of new hard-to-machine materials and requirements for tolerances of a precision that can approach one part in ten thousand, the best human operators have reached the limit of their ability. These requirements, together with a need for component flexibility, have lead to a form of automatic machine control known by the generic name *numerical control* (frequently abbreviated NC in this text and in engineering literature generally).

The history of NC began in the late 1940s, when John T. Parsons proposed

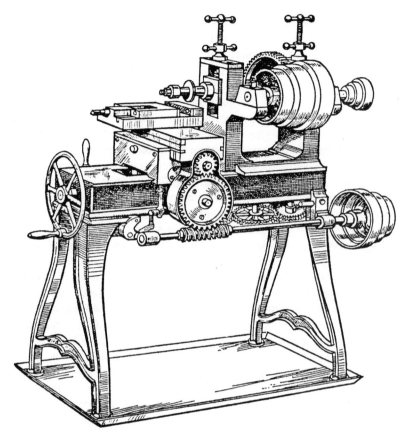

Figure 1.1 The Lincoln Miller—an early machine tool. (From Rolt, L. T. C., *A Short History of Machine Tools*, copyright © 1965 by M.I.T. Press. Reproduced with permission of B. T. Batsford, Ltd., London.)

that a method of automatic machine control be developed which would guide a milling cutter to generate a smooth curve. As he conceived it, the coordinate points would be coded onto punched cards fed to a machine controller which would cause a modified milling machine to move in small incremental steps to achieve the desired path. In 1949 the U.S. Air Force commissioned the Servomechanisms Laboratory at the Massachusetts Institute of Technology to develop a workable NC system based on Parsons' concept.

Scientists and engineers at M.I.T. selected perforated paper tape as the communication medium and initially built a two-axis point-to-point system which positioned the drilling head over the coordinate. Later, a more sophisticated continuous path milling machine was produced. Independent machine tool builders have subsequently developed the systems currently available.

By 1957, the first successful NC installations were being used in production; however, many users were experiencing difficulty in generating part programs for input to the machine controller. To remedy this situation, M.I.T. began the development of a computer based part programming *language* called APT—automatically programmed tools. The objective was to devise a symbolic language which would enable the part programmer to specify mathematical relationships in a straightforward manner (see Chapter Seven). In 1962 the first APT programming system was released for general industrial applications.

The development of numerical control technology has taken place on two major fronts. *Hardware* development concentrated on improved control systems and machine tools. Sophisticated NC machines and control systems were available by 1965 for every major machine tool configuration. *Software* development concentrated on improvements to the APT language as well as the origination of other NC programming systems.

A change in overall philosophy began in the 1970s, and numerical control was then viewed as part of a larger concept—*computer-aided manufacturing* (CAM). CAM encompasses not only NC but production control and monitoring, materials management, and scheduling. The emphasis on the use of computers in the manufacturing process has spawned new forms of numerical control: CNC (*computer numerical control*) and DNC (*direct numerical control*). Numerical control continues to develop, and possibilities that were once considered science fiction are now seen as attainable goals.

Numerical control performs best where other forms of specialized automation fail. NC is a system that can *interpret* a set of prerecorded instructions in some symbolic format; it can cause the controlled machine to execute the instructions, and then can *monitor* the results so that the required precision and function are maintained.

Numerical control is *not* a kind of machine tool but a technique for controlling a wide variety of machines. For this reason NC has been applied to assembly machines, inspection equipment, drafting machines, typesetters, woodworking machines (to name only a few applications) as well as metal cutting machine tools. Because NC was originally developed for the machine tool, and because NC metal cutting tools comprise the vast majority of all NC applications, we consider the NC machine tool exclusively in this text.

The numerical control system forms a communication link which has many similarities to conventional processes. A simplified schematic of the most important NC system elements is shown in Figure 1.2. Symbolic instructions are input to an electronic control unit which decodes them, performs any logical operations required, and outputs precise instructions that control the operation of the machine. Many NC systems contain sensing devices that transmit machine status back to the control unit. It is this *feedback* that enables the controller to verify that the machine operation conforms to the symbolic input instructions.

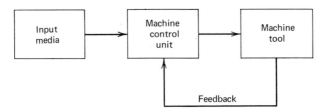

Figure 1.2 A simplified schematic of an NC system.

1.3 The Numerically Controlled Machine

We have defined NC as a method of automatic control that uses symbolically coded instructions to cause the machine to perform a specific series of operations. The most important of these instructions is that of *position*.

1.3.1 Positioning

Although modern NC systems perform many functions, the most important controlled operation is static or dynamic *positioning* of the cutting tool with the use of a system of coordinates that is general enough to define any geometric motion.

The right-handed Cartesian coordinate system, illustrated in Figure 1.3, provides a simple method for the definition of any point in three-dimensional space. While all NC machines make use of a coordinate system, some require only two axis (x and y) motion, and others require three-dimensional linear and angular axes. For the present purposes the 3-D Cartesian system is used.

Referring to Figure 1.3, the point P_1 is defined by the coordinate set $(1, 2, 2)$, where each planar subdivision represents one unit. Likewise, point P_2 is defined by the coordinate set $(3, 3, 4)$. The number of subdivisions within the three-dimensional space may be increased so that any point within the predefined boundaries may be described by a countable number of units.

Unlike a pure Cartesian system in which an infinite number of axial subdivisions is assumed, an NC coordinate system considers only a finite number of subdivisions. A later chapter will show that each *command* which is output by the control unit to the machine corresponds to one unit of motion. An NC machine is only as accurate as this smallest predefined subdivision. To illustrate this concept, consider a one-axis NC device which generates motion along the x-axis bounded by the values $x = -0.1$ m and $x = +0.1$ m. If the axis has 1000 predefined subdivisions, the machine could position to 0.2 mm. If 10,000 subdivisions were used, accuracy would increase tenfold to 0.02 mm. In these illustrations each command would correspond to 0.2 mm or 0.02 mm of motion.

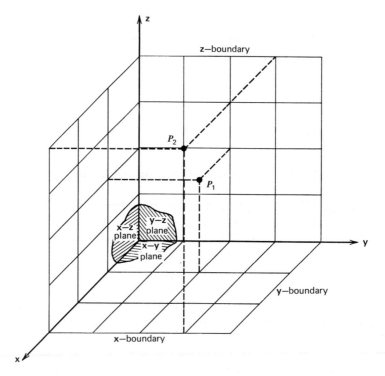

Figure 1.3 A bounded cartesian reference system.

An NC machine tool uses the coordinate system as a framework for positioning the cutting tool with respect to the workpiece. To accomplish this, points along the component profile* are defined by x, y, z coordinates. These coordinates are then fed in sequence to the NC controller which generates the appropriate positioning commands. A typical machine-axis configuration is illustrated in Figure 1.4. Any combination of spindle motion and/or work-holding table motion enables the proper position to be attained.

Positioning can be accomplished using two distinct methods. The first method, called *absolute positioning*, fixes the reference system and enables the actual x, y, z coordinates to be specified with reference to a fixed origin. Using an absolute positioning system, the points P_1 and P_2 would be specified as (x_1, y_1, z_1) followed by (x_2, y_2, z_2), regardless of the cutter position before the command is issued.

The second method of positioning uses *incremental* movement to obtain the same result. In an incremental device, the reference system is relative to the

---

* In many cases *offset* points are required (see Chapter Seven).

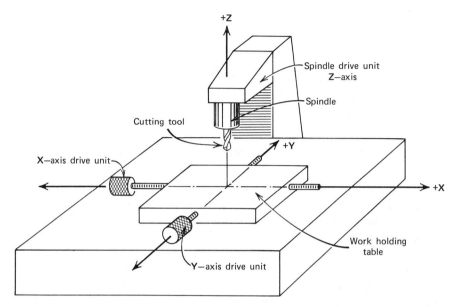

Figure 1.4 A typical NC machine axis system.

last position. For example, assume that the tool's current position is point $P_0(x_0, y_0, z_0)$. To specify a new position at P_1, the incremental distances, rather than the absolute coordinate values, are given. Let $\Delta x_{10} = (x_1 - x_0)$, $\Delta y_{10} = (y_1 - y_0)$, $\Delta z_{10} = (z_1 - z_0)$. Then, the new position is obtained by $(\Delta x_{10}, \Delta y_{10}, \Delta z_{10})$, where the reference system is assumed to have its origin at P_0. Similarly, a move from P_1 to P_2 is obtained by $(\Delta x_{21}, \Delta y_{21}, \Delta z_{21})$, where $\Delta x_{21} = (x_2 - x_1)$, etc. The incremental system therefore uses the change in x, y, z dimensions to specify position, whereas the absolute system uses coordinate values.

1.3.2 Control System

In our discussion of positioning, the path taken between points was disregarded. The path which the cutting tool follows as it traverses from point to point depends upon the type of control system used. Three basic path control systems are found in general usage.

The *point-to-point* system (also called a *positioning system*) effectively disregards the path between points. Each axis of motion is controlled independently so that the path steps from the start position to the next position as shown in Figure 1.5. The path shown is not unique as some point-to-point systems first satisfy the x command and then the y, whereas others reverse the

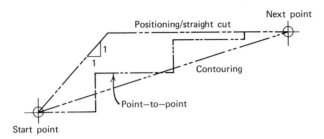

Figure 1.5 Comparison of control system paths.

order of execution. Because the traverse path is not controlled, point-to-point systems can only be used in NC applications in which a discrete operation, such as drilling a hole, retracting, drilling another hole, occurs at a given stationary location.

Positioning/straight-cut systems provide a somewhat greater degree of axis coordination than point-to-point devices. The straight-cut system has the ability to accurately follow straight paths along each machine axis. Many controllers of this type also produce limited diagonal paths by maintaining a one-to-one relationship among the motions of each axis.

Contouring systems are the most versatile and sophisticated NC devices. The contouring controller generates a path between points by *interpolating* intermediate coordinates. All contouring systems have a linear interpolation capability (i.e., the ability to generate a straight line between two points). Many systems have additional interpolation capabilities, and because the contouring system provides a predictable, accurate path between points, any path in space can be traced.

Figure 1.5 compares the typical motion for each system. The change in position with respect to time must also be considered for successful machining. This and other topics concerning the above systems are discussed in Chapters Three and Seven.

1.3.3 Communication Media

A numerically controlled machine will function only if the proper instructions are developed and passed to the machine control. The process by which the symbolic NC instructions are transferred to the control unit is termed the *communication cycle*. The cycle begins with the development of a set of NC instructions, called a *part program*, that specifies positioning data and related machining functions in a machine-readable format.

The next step in the communication cycle is the physical transfer of the part program to the machine controller. The *communication medium* transports a symbolically coded part program to the control unit. Even a relatively short set of NC instructions may contain hundreds, and possibly thousands, of alphanumeric characters and special symbols. For this reason a communication medium must represent a symbolic code in a compact form which can be easily deciphered by the machine control. The various NC communication media are discussed in Chapter Six.

1.3.4 NC Machine Configurations

A large variety of NC machine tools is currently available. Figure 1.6 illustrates three machine configurations which exhibit increasing levels of sophistication, cost, and capability. The NC drilling machine (Figure 1.6a) represents a simple two-axis point-to-point NC device. The lathe (Figure 1.6b) provides continuous path motion in the axial (z) and radial (x) directions. The rotary axis (c) may

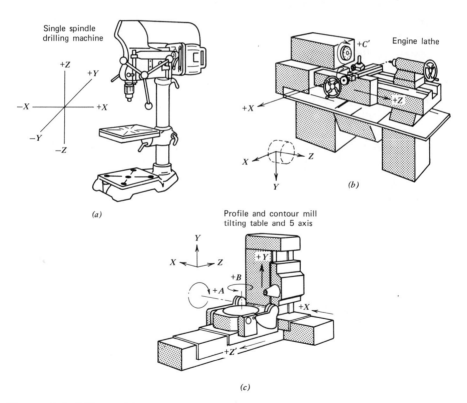

Figure 1.6 NC machine tool configurations as published by the *Electronic Industries Association*, EIA standard RS-267.

also be controllable. The profile and contour mill illustrated in Figure 1.6c reflects a high degree of sophistication by maintaining five axes of motion (three linear and two angular). With the addition of an automatic tool changing device it can perform multiple machining operations on complex components. Although many other machine configurations exist (Reference 1), our discussions in this book are concerned with tools that are functionally similar to those described.

1.4 Numerical Control Applications

As with other expanding technologies, there is a tendency to consider numerical control as a final solution to a broad range of manufacturing problems; however, NC application in certain manufacturing situations would be highly undesirable. Figure 1.7 illustrates the appropriate application areas for NC, based on the criteria of number of parts to be produced and their complexity.

An NC machine is most efficiently used in an environment that takes advantage of the inherent flexibility of NC. The precise level of control attributed to a numerically controlled device enables it to perform complex operations often beyond the capability of a human operator. For these reasons, numerical control is best suited to relatively low volume runs of complex and varied components. However, NC can also be used to produce large numbers of complex components and/or small numbers of simple ones.

The application areas of general purpose conventional machines and special purpose automated equipment are also illustrated in Figure 1.7. Dashed lines separating the application areas represent general boundaries only. Depending

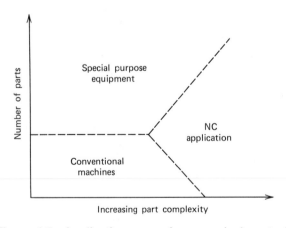

Figure 1.7 Application zones for numerical control.

on manufacturing requirements and available equipment, significant variations can occur.

Specific application areas for numerical control range from the manufacture of knitted fabrics to the fabrication of structural members for jet aircraft. In the following sections we present a brief overview of some specific applications in which NC has been utilized. An excellent in-depth discussion of many NC applications in industry can be found in Reference 2.

1.4.1 Metal-Cutting Machine Tools

Numerical control was introduced and developed in the metalworking industry, and the largest concentration of NC equipment remains in metalworking shops. NC has been successfully implemented for milling, drilling, grinding, boring, punching, turning, sawing, and routing machines. It is interesting to note that NC has made possible the development of machines with basic capabilities that far surpass their conventional counterparts. For example, sophisticated NC milling machines maintain control over five axes of motion. Such devices can literally *sculpt* complex undulating surfaces which would be impossible to machine manually.

A new breed of NC machine tool, called the *machining center*, incorporates the functions of many machines into a single device, thereby reducing work handling. It can access multiple tools to perform such operations as milling, drilling, boring, and tapping.

1.4.2 Automated Drafting

Automated drafting equipment is generally part of a larger *computer-aided design* (CAD) network. CAD systems, making use of both large-scale and minicomputers, are proliferating in the industry. Numerical control is used to translate a blueprint, drawing, or graph, which has been converted to digital format, to pen motions on an automatic drafting machine.

Effectively, an NC drafting machine is a two-axis contouring device which controls a pen or stylus as it moves over drawing paper or film. Data is supplied to the drafting machine control unit through a variety of communication media. Computer generated punched or magnetic tape is the most common mode of communication.

Although the size, accuracy, and configuration of NC drafting equipment varies among manufacturers, it is not uncommon to find machines capable of producing drawings 1.5 m × 2 m in size with an accuracy of 0.02 mm. Many sophisticated systems incorporate a digitizing feature which enables a machine

Figure 1.8 A computer-aided design (CAD) system (*courtesy Gerber Scientific Co.*).

operator to generate computer-compatible data from existing drawings. A typical CAD system is illustrated in Figure 1.8.

1.4.3 Electronics Assembly

The electronics industry uses NC machines for wire wrapping, filament winding, and circuit board operations. NC is also applied in the microelectronics field to aid in the design and manufacture of integrated circuitry. To illustrate the use of NC in these application areas, two widely used forms of numerical control are considered.

Because of the continuing miniaturization of electronic components, the need for precise NC positioning devices in the manufacture of printed circuit boards has increased. A circuit board with a hole pattern of 0.2 mm is not uncommon. Multiple spindle, point-to-point NC drilling machines are used for drilling patterns often containing thousands of holes.

Possibly the most widely used NC equipment in the electronics industry is the wiring machine. This device is programmed to position the wire above the appropriate pin, automatically strip the wire, and insert it into the pin, thereby effecting the connection. Hundreds of NC wiring machines throughout the industry have made possible improved efficiency in the manufacture of a wide range of electronic devices.

1.4.4 Quality Control/Inspection

A numerically controlled inspection machine uses its precise positioning ability to record the deviation between the actual component and the required surface. The following quality control problem illustrates this concept (Reference 2). A large parabolic surface is to be fabricated from smaller sections as illustrated in Figure 1.9. Each panel must be inspected to assure that it does not deviate from a paraboloid. Tolerance requires that 100 points per section be verified.

The equation for each paraboloid section is known, allowing the ideal values for each inspection point to be determined. These coordinates are then programmed for use by the NC inspection machine. As the NC machine moves from point to point on the surface, a linear displacement transducer follows the actual surface contour as shown in Figure 1.9. If the programmed surface coordinate (i.e., the true paraboloid coordinate value) and the actual surface coordinate agree, the transducer outputs a zero signal. If the actual and ideal coordinates are not the same, the transducer outputs a signal that records the grid point number and the deviation. An inspection machine shown in Figure 1.10 has an axis configuration similar to that in Figure 1.4. A transducer replaces the cutting tool.

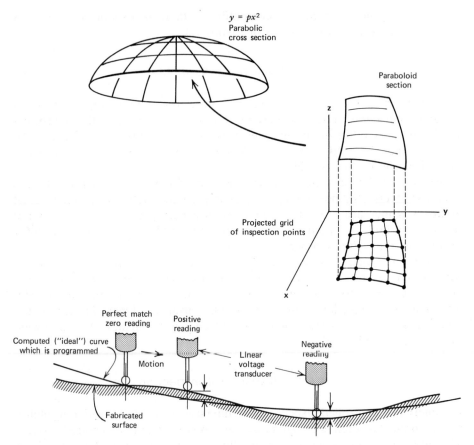

Figure 1.9 Schematic for NC inspection of a parabolic surface. (From Olesten, N.O., *Numerical Control*, copyright © 1970 by John Wiley & Sons. Reproduced with permission.)

1.4.5 Other Numerical Control Applications

A potential numerical-control application exists whenever precision positioning operations are encountered. The following paragraphs discuss some of the less familiar (although not less important) application areas. As the field of *programmable automation* continues to evolve, NC will be used increasingly in all phases of the manufacturing process.

NC welding equipment is used to provide uniformity and accuracy in a process that is affected by many external variables. By maintaining proper electrode to material distances, feed angles, feed rates, and position accuracy,

Figure 1.10 A three axis NC inspection machine (*courtesy McDonnell Douglas Corp.*).

the NC welder can maintain critical fabrication requirements. High energy welding techniques that are capable of concentrating an electron beam or laser can be precisely controlled using numerical control.

The NC welding machine makes use of the same control principles as multiple-axis metal cutting machine tools. The controller maintains position and velocity and can be programmed in the same way as machine tools. Five-axis welding machines have been used in the aerospace industry to weld along boundaries of curved surfaces.

Numerical control applications in the *woodworking* industry are similar in many respects to those encountered in metalworking. However, the large increase in feeds and speeds which can be used to mill, route, shape, and drill wood provides the justification of NC woodworking equipment. The greatest benefits of NC in this area of application include a reduction in required lot size due to the flexibility of the equipment, reduced machining time, smoother finishes which eliminate or reduce sanding, and the ability to economically produce complex designs.

The textile industry makes use of NC machines for *knitting* and *fabric cutting.* NC knitting machines maintain control over needle placement, color pattern, and garment parameters such as shape and size. NC fabric cutting machines can cut multiple layers of fabric to precise size patterns. Both kinds of equipment use punched control tapes which provide the necessary design and manufacturing data.

In heavy industries like shipbuilding, NC is used for such diversified operations as *flame cutting, riveting,* and *tube bending.* The plastics industry has recently applied numerical control to the *injection molding process* and obtained an increase in production rates and a decrease in rejects.

1.5 Present and Future Trends

Numerical control technology has evolved through five distinct phases to its current level of sophistication. The first NC machines provided rate and motion control with few auxiliary functions. Early control units, like early computers, depended upon vacuum tubes for electronic operations, but with the advent of the transistor, more sophisticated logical functions were implemented. Circular interpolation; cutter radius, tool offset and tool radius compensation; word address control; spindle speed control; and tool changes were some of the functions that then became available. When integrated circuitry became commonplace, duplicate functions, such as absolute or incremental coordinate specification could be economically offered. The application of *mini* - and even *micro* computers brought about part program storage and source language editing facilities, as well as the first practical links to CAM. Currently, refinements in solid state technology provide more powerful NC software capabilities.

The expanding usage of programmable minicomputers will make possible NC systems that automatically compensate for machine tool mechanical inaccuracies, and provide maintenance and fault finding assistance for logical and sensory circuits. Practical application oriented adaptive control systems (Chapter Nine) now provide methods for process optimization, and future machines, currently on drawing boards, will have the inherent capability of being integrated into a hierarchical CAM structure.

References

1. *NC/CAM Guidebook, Modern Machine Shop*, vol. 47, January 1975, pp. 196–205.
2. Olesten, N. O., *Numerical Control*, Wiley, New York, 1970.

Additional Reference

Wiener, N., *Cybernetics*, 2nd ed., M.I.T. Press, Cambridge, Mass., 1961.

Chapter Two
Control System Fundamentals

Numerically controlled machines often weigh up to 100 tons and yet are required to position a cutting tool with an accuracy of the order of 0.002 mm. The control system must move the tool at feedrates as high as 8 cm/sec while encountering loads which may vary dramatically on a given path. The NC machine must have dynamic response characteristics that enable it to follow intricate contours with a minimum of path error. Clearly, these requirements dictate a *control* system that is matched to the mechanical characteristics of the machine it drives.

A control system is a combination of devices that regulate an operation by administering the flow of energy and other resources to and from that operation. Essentially, a control system is made up of interrelated subsystems that perform tasks which in orthodox machining processes are managed by an intelligent human operator.

The control systems for NC machines therefore serve to replace the human machine operator and significantly improve upon even the best human performance. A direct analogy can be made between numerical controls and the human operators they replace. Both:

1. *Sense* the current status of the machine.
2. Make *logical decisions* which are required to accomplish a task.
3. *Communicate* decisions to the machine by *actuating* proper mechanical devices.
4. Have the ability to *store information*: instructions, data, and the results of logical decisions.

In summary, a machine control system is a combination of electronic circuitry, sensing devices, and mechanical components which guide the cutting tool along a predefined path.

In the sections that follow, the reader is introduced to the theory of machine

18

control. We use the term "introduced" in its true sense, in that an in-depth discussion of control theory is beyond the scope of this book. Numerous texts (References 1–3) of varying levels of sophistication and completeness have been written on control theory, and the interested reader is urged to pursue the subject.

2.1 Control System Concepts

2.1.1 Automatic Control

We have attributed four major characteristics to any control system: the ability to *sense* data; make *logical decisions*; *communicate* and *actuate*; and *store information* in a memory. If a system performs these functions without the aid of a human operator, by definition it *automatically* controls a given task. Let us consider one such simple operation.

Suppose that an electric motor is used to drive a freight conveyor between two floors of a building. To operate smoothly the conveyor must maintain a constant speed as additional loads are added from loading shoots. We can assume that the speed of rotation of an electric motor is directly proportional to the electric current supplied to it. Hence as the load increases on the conveyor, the torque requirement of the motor is increased and additional power must be supplied to maintain constant speed. This is usually achieved by increasing the supply voltage.

If a motor speed indicator is within sight of a human operator, he can control the conveyor speed by increasing or decreasing the voltage supply manually. Such a system is illustrated schematically in Figure 2.1.

As indicated by Figure 2.1*a*, the operator first senses that the speed is below the desired value. Logically, he determines that an increase in power will increase motor speed. He communicates a decision, by actuating the power supply control lever, to increase voltage until the desired speed reading is obtained. In Figure 2.1*b* the desired speed is read; hence, the operator takes no action. If the load is decreased so that a speed greater than the desired value is registered (Figure 2.1*c*), the operator will decrease voltage.

The process described above is illustrated schematically in Figure 2.2. The system indicated in the figure is not automatic. However, if the human operator block were replaced with a device that performed the functions shown, an automatic control system would be realized.

Automatic control systems are sometimes called *servomechanisms.* A servomechanism may be a complex array of electromechanical components or it may be a simple mechanical device. Regardless of the complexity of the servomechanism, it always exhibits the features illustrated in Figure 2.2.

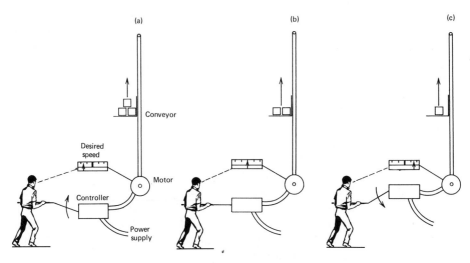

Figure 2.1 A manually controlled conveyor.

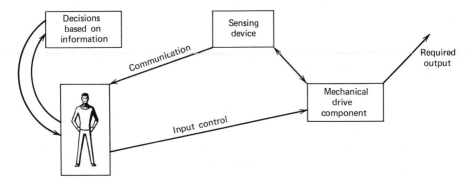

Figure 2.2 Schematic of information flow for conveyor system.

Consider a simple mechanical device that performs the automatic control function for our conveyor system (Figure 2.3). Such a device, called a Watt speed governor,* can automatically control the speed of an electric motor.

As shown in Figure 2.3, the Watt speed governor consists of two weighted arms connected to a shaft, S_2, and a sleeve. The motor shaft, S_1, is coupled to the governor shaft, S_2, so that the rotational velocity of S_2 is proportional to S_1. As motor speed increases, centrifugal force moves the weights, W, outward, causing the weight arms to rotate through some angle, θ. The weights, angle,

* The Watt speed governor was developed by James Watt (1736–1819).

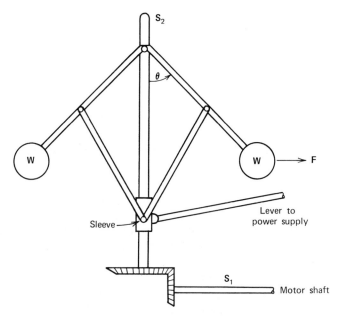

Figure 2.3 The Watt mechanism—an early example of an automatic control unit.

and sleeve position are chosen so that the proper speed will cause the voltage supply lever to be positioned at a neutral position. Variations in speed will cause the appropriate movement of the sleeve and lever.

2.1.2 Open Loop Control

An open loop control system is the simplest and least complex form of servomechanism. It is characterized as a system which lacks *feedback*; that is, once an input control signal is sent, there is no sensing device to confirm the action of the control signal.

In the open loop motor speed control system, the power supply lever would be set to a position at which the desired speed is indicated. Should the load on the conveyor be varied, the motor speed would also vary. Because there is no way of automatically sensing this speed variance (i.e., no feedback), the change in required power input must be manually compensated.

Obviously, open loop control systems can be applied only in situations in which no change in output condition occurs. Such a change is referred to as *disturbance* or *secondary input*; and for an accurately functioning open loop system, the disturbance must be zero. An open loop system can be represented schematically as illustrated in Figure 2.4.

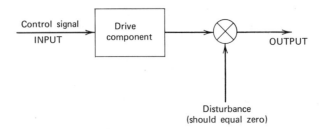

Figure 2.4 Open loop control system schematic.

2.1.3 Closed Loop Control

A closed loop control system is characterized by the presence of feedback. The input to the system and the output from the drive and/or positioning components are continually compared. Closed loop controls make use of an error detector that returns a signal which is proportional to the difference between input and feedback. Such a control loop is illustrated in Figure 2.5.

The conveyor system as modified for use with the Watt speed governor (Figures 2.1 and 2.3) is an example of a closed loop system. Differences between motor speeds are sensed, and the sleeve is automatically moved to correct the discrepancy. It is important to note that the output of a closed loop control can be made relatively independent of outside disturbances. For this reason, closed loop controls are used where a *load* of varying magnitude (involving disturbance) is present. Such systems are extremely accurate.

The differences between open and closed loop controls can be considered with respect to response and accuracy. Closed loop control systems out

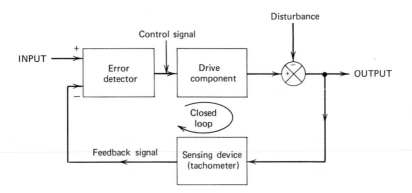

Figure 2.5 Closed loop control system schematic.

perform open loop systems in both respects. However, the very nature of feedback control can make the system *unstable*. That is, the system may tend to oscillate about the desired value instead of achieving it.

2.1.4 System Response

Every dynamic system has a *characteristic response time*; this is a measure of the time required by a system to react to a given input signal. For example, the time required to raise the temperature 5° in a typical home heating/air conditioning system might be 30 minutes, whereas the response time of a toaster might be 5° in 3 seconds.

The characteristic response is usually specified as the *time constant* of the system, where

$$\text{time constant, } \tau \equiv \text{characteristic response} \times 0.63.$$

That is, the time constant is measured as the time required for the system response to reach 63 percent of the desired value. *Bandwidth* is another term related to system response and is defined as the range of frequencies over which the system reacts favorably. At frequencies outside the system bandwidth, response attenuates rapidly. Bandwidth is also used to denote the opposite of response time. A system which has a small bandwidth generally has a long characteristic response and vice versa.

Let us examine how time response affects the operation of a closed loop system. Consider the home heating/air conditioning system. If the thermostat were changed by 5° once every 30 minutes, the system would be able to react to the input commands. If, however, the thermostat were varied by 5° increments at 5-minute intervals, the system would lag behind the command signal. Exaggerating the effect, a change of ± 5° in 1-minute intervals would result in no appreciable temperature change. It follows, therefore, that a knowledge of the frequency of command signals is essential when the system time constant is chosen. In an NC system the choice of time constants is of vital importance.

2.2 Feedback Control System Elements

The exact hardware configuration of an automatic control system is dependent upon specific system requirements. Typical NC servomechanisms require the following components: (1) transducers, (2) servomotors, and (3) control amplifiers. These system elements are integrated into a closed loop, illustrated schematically as a *block diagram* in Figure 2.6.

A transducer is defined as any device that *senses* an output condition and

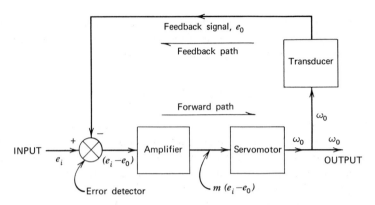

Figure 2.6 Control system block diagram.

transforms the sensed information into a form which can be understood by the servo system. A device that outputs a voltage in direct proportion to a measured linear movement is a typical example.

The input signal is generally summed with the feedback signal, thereby producing a corrected control signal to the system drive components. Control amplifiers are used to boost the magnitude of the input and feedback signals to a value that is sufficient to power the drive components. If S_i is the sum of all input signals to a control amplifier and S_o is the output signal, the *amplifier gain* is defined as

$$m = \frac{S_o}{S_i}$$

Servomotors respond to the outputs of the control amplifiers and provide rotary or linear motion required by the drive system. Many types of servomotors are available for use in NC equipment.

2.2.1 The Block Diagram

Figure 2.6 illustrates a control system block diagram. Each block of the diagram represents a device, a part of a device, or a function that occurs in the system.

In the figure, the input and output of each block are indicated. Following the loop in a counterclockwise direction, the input signal, e_i, is summed with the feedback signal, e_0, to produce input to the control amplifier. The amplifier magnifies this signal (gain $= m$) for input to the servomotor. The output of the motor is *not* a signal, but a rotary motion, ω_0, which is measured by the transducer. The output signal of this transducer, e_0, closes the loop.

The motor runs at constant output (or no output, depending on the application) provided the error signal, $e_i - e_0$, equals zero. If however, $e_0 \neq e_i$, the voltage difference will be amplified as a correction for input to the servomotor.

2.3 The Transfer Function

In a typical control system, the block diagram for the control loop would be more complex than that illustrated in Figure 2.6. Regardless of the number of blocks on a path the input to the path and the output from it have primary importance.

A *transfer function* enables all of the individual elements on a path to be lumped into a single block. This block can then be applied to the input to produce the required output.

Consider a path which has an input I and an output C. The difference between input and output is the error, E,

$$E = I - C \tag{2.1}$$

If all the elements on the path are represented by a single block, the *forward path* can be illustrated as shown in Figure 2.7.

The block Φ_f represents a transfer function. By definition*

$$\Phi_f = \frac{\text{output}}{\text{input}} = \frac{C}{I} \tag{2.2}$$

Each device in the forward path has a transfer function. The overall transfer function, Φ_f, can be shown to be

$$\Phi_f = \Phi_1 \cdot \Phi_2 \cdots \Phi_n \tag{2.3}$$

provided that the individual blocks are independent.

Consider Figure 2.6. Let the forward path have a transfer function Φ_f, and the *feedback path* have the transfer function Φ_b. Referring to Figure 2.8, the closed loop transfer function

$$\Phi = \frac{\text{output}}{\text{input}} = \frac{C}{I}$$

Figure 2.7 Transfer function block.

---

*We shall see later that C and I are *operational* functions.

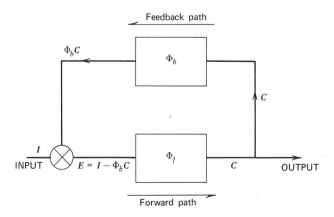

Figure 2.8 Schematic for the development of a closed loop transfer function.

However

$$\Phi_f = \frac{C}{I - \Phi_b C}$$

or solving for the output

$$C = \frac{\Phi_f I}{1 + \Phi_f \Phi_b}$$

Therefore, the closed loop transfer function

$$\Phi = \frac{\Phi_f}{1 + \Phi_f \Phi_b} \qquad (2.4)$$

If the feedback path is broken before it enters the error detector, an open loop would be created. The open loop transfer function is defined as the ratio of the feedback signal to the error signal,

$$\text{open loop transfer function} = \frac{\Phi_b C}{E} = \Phi_f \Phi_b$$

In our system $\Phi_b = 1$, since input and feedback signals are compared directly. Therefore the closed loop transfer function becomes

$$\Phi = \frac{\Phi_f}{1 + \Phi_f}$$

or

$$\Phi = \frac{\text{open loop transfer function}}{1 + \text{open loop transfer function}}$$

In later sections it will be shown that the transfer function is used in all phases of control system analysis because it yields information concerning system response and stability.

2.3.1 An Example of a Transfer Function

The following example illustrates the development of a transfer function, Φ_f, for the simple circuit shown.

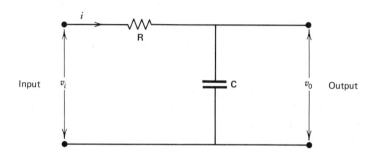

From elementary electronic circuit analysis, Kirchoff's second law can be written

$$v_i = iR + \frac{1}{C}\int_0^t i\,dt \tag{i}$$

and

$$v_0 = \frac{1}{C}\int_0^t i\,dt$$

therefore

$$i = C\frac{dv_0}{dt} \tag{ii}$$

Substituting equation (ii) into equation (i)

$$v_i = RC\frac{dv_0}{dt} + v_0 \tag{iii}$$

If we let **d** be an operator which represents d/dt, then equation (iii) can be written

$$v_i = v_0\,(\mathbf{d}\,RC + 1)$$

From our definition of a transfer function

$$\Phi = \frac{v_0}{v_i} = \frac{1}{\mathbf{d}\,RC + 1} \tag{iv}$$

Hence, equation (iv) is the transfer function of the R-C circuit.

2.4 Introduction to the Mathematics of Control

2.4.1 Control System Equations

The response of any dynamic system (automatic or manual) is often specified as a function of time. Closed loop control systems can be represented by a differential equation of the form

$$A_n \frac{d^n x_r}{dt^n} + A_{n-1} \frac{d^{n-1} x_r}{dt^{n-1}} + \cdots + A_0 x_r = B_m \frac{d^m x_f}{dt^m} + B_{m-1} \frac{d^{m-1} x_f}{dt^{m-1}} + \cdots + B_0 x_f$$

where x_r is the output response variable, and x_f is the input forcing function. Generally an automatic control system is modeled on a second order system, even though actual systems are of a higher order. This approximated representation achieves reasonable accuracy. Hence, in the preceding general expression, $n = 2$. It should also be noted that for an automatic control system, $n \geqslant m$; that is, the system output should never increase without bound.

The following discussion considers only systems with a transfer function invariant with respect to the system input. Such systems are called *linear*.

2.4.2 Operator Notation—The Laplace Transform

The solution of a second order linear differential equation of the form

$$A_2 \frac{d^2 x}{dt^2} + A_1 \frac{dx}{dt} + A_0 x = f(t)$$

can be performed using classical methods. For a complete discussion of these solutions, the reader is directed to References 4–6.

In the analysis of control systems, classical solution methods are rarely used. Generally, a control engineer *transforms* the differential equation into a more easily evaluated algebraic equation. He does this using an operator function called the *Laplace transform*.

The Laplace transform is defined in the following manner:

$$F(s) = \int_0^\infty f(t) e^{-st} \, dt \qquad (2.5)$$

where s is a complex number of the form $\sigma + j\omega$. The transform of a function, $f(t)$, becomes the function in s, $F(s)$. The analysis of differential equations using the Laplace transform rarely requires the use of equation (2.5), since extensive tables for a myriad of functions are available. An abridged version is given in Table 2.1.

Table 2.1
A Brief Table of Laplace Transforms

$F(s)$	$f(t)$
$\dfrac{1}{s}$	1
$\dfrac{1}{s^2}$	t
$n!/s^{n+1}$	t^n
$\dfrac{1}{s+a}$	e^{-at}
$\dfrac{1}{(s+a)^2}$	te^{-at}
$\omega/(s^2+\omega^2)$	$\sin \omega t$
$s/(s^2+\omega^2)$	$\cos \omega t$
$\omega/(s^2-\omega^2)$	$\sinh \omega t$
$s/(s^2-\omega^2)$	$\cosh \omega t$
$\omega/[(s+a)^2+\omega^2]$	$e^{-at}\sin \omega t$
$(s+a)/[(s+a)^2+\omega^2]$	$e^{-at}\cos \omega t$
$\dfrac{1}{s^2+2\zeta\omega s+\omega^2}$	$\dfrac{1}{\sqrt{1-\zeta^2}}e^{-\zeta\omega t}\sin \omega \sqrt{1-\zeta^2}\,t$
$sF(s)-f(0)$	$f'(t)$
$s^2F(s)-sf(0)-f'(0)$	$f''(t)$
$s^nF(s)-s^{n-1}f(0)-s^{n-2}f'(0)-$ $\cdots -f^{(n-1)}(0)$	$f^{(n)}(t)$

To illustrate, let us examine the Laplace transformation, $\mathscr{L}f(t)$, of the expression e^{-at}, where a is a constant.

$$F(t) = e^{-at}$$

and from equation (2.5)

$$F(s) = \int_0^\infty e^{-at}e^{-st}\,dt$$

$$F(s) = \left[\frac{e^{-(s+a)t}}{-(s+a)}\right]_0^\infty$$

Therefore

$$F(s) = \mathscr{L}(e^{-at}) = \frac{1}{s+a}$$

The above expression corresponds exactly to the fourth entry in Table 2.1. All entries were derived in an analogous manner.

Just as every expression $F(t)$ has a Laplace transform $F(s)$, each expression in s (even if it is not in the tables) has an *inverse Laplace transform* in t. Hence

$$f(t) = \mathcal{L}^{-1}F(s)$$

For example, from Table 2.1

$$\mathcal{L}^{-1}\left\{\frac{1}{(s+a)^2}\right\} = te^{-at}$$

2.4.3 An Example of the Use of Tabularized Transforms

From the table of Laplace transforms

$$\mathcal{L}f'(t) = s\mathcal{L}f(t) - f(0) \tag{i}$$

and

$$\mathcal{L}f''(t) = s^2\mathcal{L}f(t) - sf(0) - f'(0) \tag{ii}$$

Using equation (i) we will demonstrate that

$$\text{(a)} \qquad \mathcal{L}e^{at} = \frac{1}{s-a}$$

$$\text{(b)} \qquad \mathcal{L}\sin\omega t = \frac{\omega}{s^2 + \omega^2}$$

(a) Since

$$f(t) = e^{at}, \text{ then } f'(t) = ae^{at} \text{ and } f(0) = 1, f'(0) = a$$

Using equation (i)

$$\mathcal{L}(ae^{at}) = s\mathcal{L}e^{at} - 1$$

or removing the constant

$$a\mathcal{L}e^{at} = s\mathcal{L}e^{at} - 1$$

Rearranging

$$\mathcal{L}e^{at} = \frac{1}{s-a}$$

(b) Since $f(t) = \sin\omega t$, then $f'(t) = \omega\cos\omega t$ and $f''(t) = -\omega^2\sin\omega t$. Given the initial conditions

$$f(0) = 0, \text{ and } f'(0) = \omega$$

Using equation (ii)

$$\mathcal{L}\{-\omega^2\sin\omega t\} = s^2\mathcal{L}\sin\omega t - \omega$$

or

$$-\omega^2\mathcal{L}\sin\omega t = s^2\mathcal{L}\sin\omega t - \omega$$

Then

$$\mathcal{L}\sin\omega t = \frac{\omega}{s^2 - \omega^2}$$

2.4.4 An Example to Illustrate the Solution of a Second Order Equation

Given the initial conditions $y(0) = 10$ and $y'(0) = 0$, we are required to solve the following differential equation using Laplace transforms:

$$\frac{d^2y}{dt^2} + 2y = 0 \tag{i}$$

Taking the Laplace transform of equation (i),

$$s^2\mathbf{y} - 10s + 2\mathbf{y} = 0$$

or

$$\mathbf{y} = \frac{10s}{s^2 + 2}$$

where $\mathbf{y}$ is the transformed y value.

Now, the inverse transform

$$y = \mathcal{L}^{-1}\mathbf{y} = 10\mathcal{L}^{-1}\left\{\frac{s}{s^2 + 2}\right\}$$

From Table 2.1

$$\mathcal{L}^{-1}\left\{\frac{s}{s^2 + 2}\right\} = \cos\sqrt{2}t$$

Therefore

$$y = 10\cos\sqrt{2}t$$

which is the time dependent solution of equation (i).

2.4.5 Partial Fraction Expansion

Evaluation of the inverse Laplace transform often involves an algebraic expression in s which cannot be found in the transform table. The *method of partial fractions* provides a means for reducing an expression of the form

$$F(s) = \frac{N(s)}{(s + r_1)(s + r_2)(s + r_3)\cdots(s + r_n)} = \frac{N(s)}{D(s)} \tag{i}$$

to the equivalent expression

$$F(s) = \frac{C_1}{s + r_1} + \frac{C_2}{s + r_2} + \frac{C_3}{s + r_3} + \cdots + \frac{C_n}{s + r_n} \tag{ii}$$

where each term on the right side of the above equation has a known inverse transform.

To determine the nth constant, C_n, both sides of the equation are multiplied by the denominator of the nth term, that is, $(s + r_n)$. The Laplace variable s is then replaced by $-r_n$ in each of the remaining terms. Hence,

$$C_n = \lim_{s \to -r_n} \left[\frac{N(s)}{D(s)} \cdot (s + r_n) \right] \qquad \text{(iii)}$$

For example, to calculate the constant C_3 in equation (ii)

$$C_3 = \frac{N(s)}{D(s)} \cdot (s + r_3) \big|_{s = -r_3} \qquad \text{(iv)}$$

If equation (i) takes a form that has repeated roots in the denominator

$$F(s) = \frac{N(s)}{D(s)} = \frac{N(s)}{(s + r_1)^m (s + r_{m+1})(s + r_{m+2}) \cdots (s + r_{m+n})}$$

Then an equivalent expression can be written,

$$F(s) = \frac{C_1}{(s + r_1)^m} + \frac{C_2}{(s + r_1)^{m-1}} + \cdots + \frac{C_m}{s + r_1} + \frac{C_{m+1}}{s + r_{m+1}} + \cdots + \frac{C_{m+n}}{s + r_{m+n}} \qquad \text{(v)}$$

It can be shown (Reference 3) that the constants in equation (v) may be determined using the following relationships:

$$C_1 = \frac{N(s)}{D(s)} \cdot (s + r_1)^m \big|_{s = -r_1} \qquad \text{(vi a)}$$

$$C_2 = \frac{d}{ds} \left[\frac{N(s)}{D(s)} \cdot (s + r_1)^m \right]_{s = -r_1} \qquad \text{(vi b)}$$

$$C_m = \frac{1}{(m-1)!} \frac{d^{m-1}}{ds^{m-1}} \left[\frac{N(s)}{D(s)} \cdot (s + r_1)^m \right]_{s = -r_1} \qquad \text{(vi c)}$$

The constants C_{m+1} through C_{m+n} are determined using the method expressed in equation (iii).

2.4.6 Example—The Use of Partial Fractions

We are required to expand the following functions of s in partial fraction representations:

$$\text{(a)} \qquad F(s) = \frac{(s+3)}{s(s^2 + 8s + 7)}$$

$$\text{(b)} \qquad F(s) = \frac{(s+1)}{s(s+3)^2(s+2)}$$

(a) Factoring the denominator

$$F(s) = \frac{N(s)}{D(s)} = \frac{s+3}{s(s+7)(s+1)}$$

and

$$F(s) = \frac{C_1}{s} + \frac{C_2}{s+7} + \frac{C_3}{s+1}$$

Using equation (iii) in Section 2.4.5, the constants can be evaluated:

$$C_n = \frac{N(s)}{D(s)} \cdot (s + r_n)|_{s=-r_n}$$

Therefore

$$C_1 = \frac{(s+3)}{s(s+7)(s+1)} \cdot s|_{s=0} = \frac{3}{7}$$

$$C_2 = \frac{(s+3)}{s(s+1)}\bigg|_{s=-7} = \frac{-4}{-42} = \frac{2}{21}$$

$$C_3 = \frac{(s+3)}{s(s+7)}\bigg|_{s=-1} = \frac{2}{-6} = -\frac{1}{3}$$

Hence

$$F(s) = \frac{3}{7s} + \frac{2}{21(s+7)} - \frac{1}{3(s+1)}$$

The inverse Laplace transform can be applied to the above expression to yield (from Table 2.1)

$$f(t) = \frac{3}{7} + \frac{2}{21} e^{-7t} - \frac{1}{3} e^{-t}$$

(b) Rewriting $F(s)$ in partial fraction form

$$F(s) = \frac{C_1}{(s+3)^2} + \frac{C_2}{(s+3)} + \frac{C_3}{s} + \frac{C_4}{s+2}$$

From equations (vi)

$$C_1 = \frac{(s+1)}{s(s+2)(s+3)^2} \cdot (s+3)^2|_{s=-3} = -\frac{2}{3}$$

$$C_2 = \frac{d}{ds}\left[\frac{(s+1)}{s(s+2)}\right]_{s=-3} = \frac{s(s+2) - 2(s+1)^2}{s^2(s+2)^2}\bigg|_{s=-3}$$

$$C_2 = -\frac{5}{9}$$

The remaining constants are calculated using equation (iii):

$$C_3 = \frac{s+1}{(s+3)^2(s+2)}\bigg|_{s=0} = \frac{1}{18}$$

$$C_4 = \frac{s+1}{s(s+3)}\bigg|_{s=-2} = \frac{1}{2}$$

Therefore

$$F(s) = -\frac{2}{3(s+3)^2} - \frac{5}{9(s+3)} + \frac{1}{18s} + \frac{1}{2(s+2)}$$

The inverse Laplace transform can be applied to the above expression to yield

$$f(t) = -\frac{2}{3}\,te^{-3t} - \frac{5}{9}\,e^{-3t} + \frac{1}{2}\,e^{-2t} + \frac{1}{18}$$

2.4.7 The Transfer Function and Operational Expressions

Laplace transforms are used to solve equations of the general form:

$$A_n\frac{d^nc}{dt^n} + \cdots + A_1\frac{dc}{dt} + A_0c = B_m\frac{d^mi}{dt^m} + \cdots + B_1\frac{di}{dt} + B_0i \tag{2.6}$$

If zero initial conditions are assumed,* equation (2.6) can be written as

$$(A_ns^n + \cdots + A_1s + A_0)c(s) = (B_ms^m + \cdots + B_1s + B_0)i(s) \tag{2.7}$$

We have already defined the transfer function

$$\Phi_f = \frac{\text{output}}{\text{input}} = \frac{C}{I} \tag{2.2}$$

Now let us define the transfer function, $G(s)$, in terms of equation (2.7) as

$$G(s) = \frac{c(s)}{i(s)} = \frac{(B_ms^m + \cdots + B_1s + B_0)}{(A_ns^n + \cdots + A_1s + A_0)} \tag{2.8}$$

From equation (2.8) it is evident that the transfer function of a system is the ratio of two polynomials of degree m and n, respectively. $G(s)$ is of primary importance in the analysis of control systems. Not only does it enable determination of system response, but it also yields important information on the stability of the system.

2.4.8 Control System Transfer Functions

The transfer function for the control system of an NC machine is a combination of the transfer functions of the individual transducers, amplifiers, and servo elements in the loop. The following method enables individual element transfer functions to be combined into a single system transfer function.

---

*In the following development we assume initial conditions to be zero because the transfer function must be independent of initial conditions; that is, it is a property of the elements of the system only.

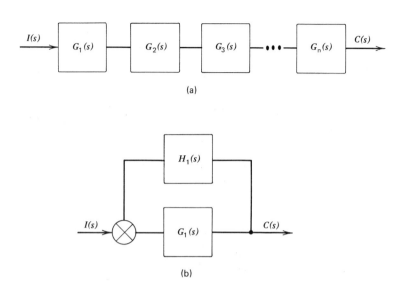

Figure 2.9 System elements in (a) cascade and (b) loop form.

In Section 2.3, equations governing the combination of individual transfer functions were presented. Equation (2.3) expressed the transfer function of a group of *cascaded* system elements (Figure 2.9a). Thus,

$$G(s) = G_1(s)G_2(s)G_3(s) \cdots G_n(s) \tag{2.9}$$

Similarly, the transfer function of a loop (Figure 2.9b) was shown to be

$$G(s) = \frac{G_1(s)}{1 + G_1(s)H_1(s)} \tag{2.10}$$

Using these expressions, a multiple loop system can be reduced to a single system transfer function.

Figure 2.10a illustrates the block diagram of a multiloop control system. This configuration is similar in many respects to the block diagram of a typical NC control system.

As indicated in Figure 2.10b–d, equations (2.9) and (2.10) are used to reduce the system to a *single transfer function*, and the expression given in Figure 2.10d is that system's transfer function. Because each element transfer function, $G_k(s)$ and $H_j(s)$, can be a polynomial expression in s, the algebraic complexity of the end product becomes significant.

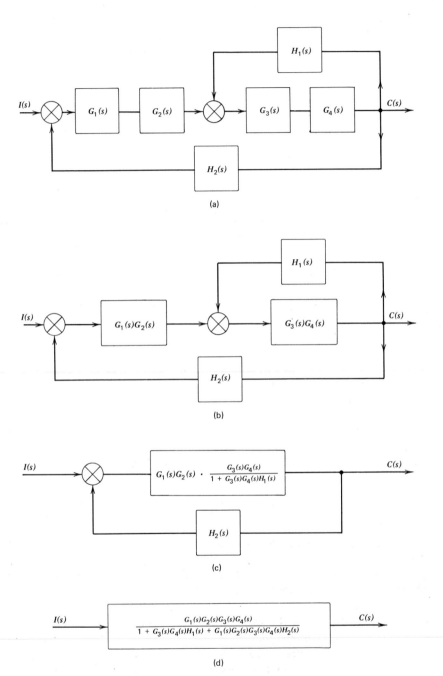

Figure 2.10 Reduction of a multiloop block diagram to a single block.

2.5 System Stability

When supplied with an input command signal, a control system must *always* tend toward a *quiescent state* (i.e., a state in which the feedback error signal is zero). Such a system may be considered to be *stable*. Consider the following dynamic response example problem.

2.5.1 An Example of Dynamic Response

Differential equations of the form

$$\frac{d^2x}{dt^2} + 2\alpha\frac{dx}{dt} + \beta^2 x = f(t) \tag{i}$$

are encountered in many engineering applications. Using Laplace transform methods, system stability can be examined for various values of α and β. The function, $x = g(t)$, can be found for the following conditions:

$$x(0) = X_0$$
$$x'(0) = V_0$$

and

$$f(t) = 0$$

Thus by applying the transforms for first and second derivatives to equation (i)

$$s^2\mathbf{x} - sX_0 - V_0 + 2\alpha(s\mathbf{x} - X_0) + \beta\mathbf{x} = 0$$

and rearranging,

$$\mathbf{x} = \frac{(s + \alpha)X_0 + V_0}{s^2 + 2\alpha s + \beta^2}$$

which we can rewrite as

$$\mathbf{x} = \frac{(s + \alpha)X_0}{(s + \alpha)^2 + \beta^2 + \alpha^2} + \frac{V_0 + \alpha X_0}{(s + \alpha)^2 + \beta^2 + \alpha^2} \tag{ii}$$

From Table 2.1, the inverse Laplace transform of equation (ii) is dependent on the value of α^2 and β^2. This value may take three forms:

1. $\beta^2 - \alpha^2 > 0$
2. $\beta^2 - \alpha^2 = 0$
3. $\beta^2 - \alpha^2 < 0$

We can examine the response for each case:

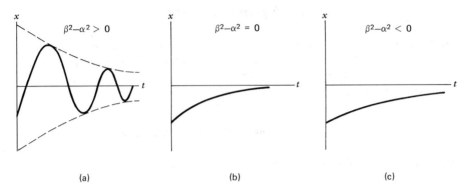

Figure 2.11 Dynamic response for (a) underdamped, (b) critically damped, and (c) overdamped systems.

Case 1. $\beta^2 - \alpha^2 > 0$

Using Table 2.1, the inverse Laplace transform for equation (ii) becomes

$$\mathcal{L}^{-1}x = X_0 e^{-\alpha t} \cos \sqrt{\beta^2 - \alpha^2}\, t \frac{V_0 + \alpha X_0}{\sqrt{\beta^2 - \alpha^2}} e^{-\alpha t} \sin \sqrt{\beta^2 - \alpha^2}\, t \qquad \text{(iii)}$$

The response indicated by equation (ii) is called *damped oscillatory* and is illustrated in Figure 2.11a. The system oscillates about a *steady state* condition (i.e., $x = 0$) with ever-decreasing amplitude. Such systems are sometimes called *underdamped.*

Case 2. $\beta^2 - \alpha^2 = 0$

Equation (ii) becomes

$$x = \frac{(s + \alpha)X_0}{(s + \alpha)^2} + \frac{V_0 + \alpha X_0}{(s + \alpha)^2} \qquad \text{(iv)}$$

and the inverse transform yields

$$x = X_0 e^{-\alpha t} + (V_0 + \alpha X_0)t\, e^{-\alpha t} \qquad \text{(v)}$$

The response indicated by equation (v) is called *critically damped* and is illustrated in Figure 2.11b. In this case the system *does not* oscillate, but instead approaches a steady state along an exponential curve.

Case 3. $\beta^2 - \alpha^2 < 0$

Equation (ii) becomes

$$x = \frac{(s + \alpha)X_0}{(s + \alpha)^2 - (\alpha^2 - \beta^2)} + \frac{V_0 + \alpha X_0}{(s + \alpha)^2 - (\alpha^2 - \beta^2)} \qquad \text{(vi)}$$

With the use of more complete Laplace transform tables (e.g., Reference 7), the inverse transform of equation (vi) can be shown to be

$$x = X_0 \cosh \sqrt{\alpha^2 + \beta^2}\, t + \frac{V_0 + \alpha X_0}{\sqrt{\alpha^2 - \beta^2}} \sinh \sqrt{\alpha^2 - \beta^2}\, t \qquad \text{(vii)}$$

The response indicated by equation (vii) is *underdamped* and is illustrated in Figure 2.11c. In this case the system reaches the steady state condition only after a long time delay.

It is evident from the preceding example that the judicious choice of the α and β values of equation (i) is quite important if the system is to behave as desired. When an automatic control system is modeled by an equation similar to (i), α and β determine the *stability* of the system.

2.5.2 Determination of Stability Using the Root Locus Technique

Methods for determining the stability of a system are available. Consider a general loop transfer function of the form:

$$G(s) = \frac{G_1(s)}{1 + H_1(s)G_1(s)}$$

The stability of the system (or element) which $G(s)$ describes is found by examining the denominator, or *characteristic*, of the expression. As $H_1(s)G_1(s)$ tends to -1, the transfer function goes to infinity and the system becomes *unstable*.

In reality, however, the characteristic takes the form

$$1 + G_1(s)H_1(s) = (s - q_1)(s - q_2)(s - q_3) \cdots (s - q_n)$$

Hence, the singularities, or *poles*, of $G(s)$ are defined as those values that make the characteristic equal to zero. An example of a pole is $s = q_1$. In a like manner, the numerator of equation (2.10) can be written

$$G_1(s) = K(s - r_1)(s - r_2) \cdots (s - r_n)$$

The *zeros* of $G(s)$ are those values of s for which the numerator of the transfer function is zero.

The poles and zeros of a transfer function may be plotted in the complex plane ($\alpha + j\omega$, where $j = \sqrt{-1}$) to provide a *root locus pattern.**

---

* The root locus technique and other stability criteria have received major attention in texts on control theory, and the reader is referred to References 1–3.

To illustrate how expression $G(s)$ can be used to indicate system stability let

$$G(s) = N\left/\left(\frac{A_1}{s-q_1} + \frac{A_2}{s-q_2} + \cdots + \frac{A_n}{s-q_n}\right)\right.$$

Assume that $q_1 = \alpha_1 + j\omega_1$ and $q_2 = \alpha_1 - j\omega_1$. The characteristic then has poles at q_1 and q_2. If the poles are plotted in the complex s-plane, points to the right of the $j\omega$-axis result, as shown in Figure 2.12a.

Rewriting the first two terms of the denominator

$$\frac{A_1}{s-(\alpha_1+j\omega_1)} + \frac{A_2}{s-(\alpha_1-j\omega_1)}$$

The inverse Laplace transform of the above expression is

$$A_1 e^{\alpha_1 t} e^{j\omega t} + A_2 e^{\alpha_1 t} e^{-j\omega t} = e^{\alpha_1 t}(A \sin \omega_1 t + B \cos \omega_1 t) \qquad (2.11)$$

Equation (2.11) indicates an exponentially increasing oscillatory response (Figure 2.12b). Since this is a diverging system response, it is undesirable because a quiescent state will not be reached. Hence, the transfer function produces instability.

Because any value of $\alpha_n + j\omega_n$ where $\alpha > 0$ yields an increasing exponential, and therefore, instability, it is seen that *any* characteristic with a pole to the right of the $j\omega$-axis indicates instability. The root locus method enables the stability of a system to be determined merely by examining the poles and zeros of the transfer function.

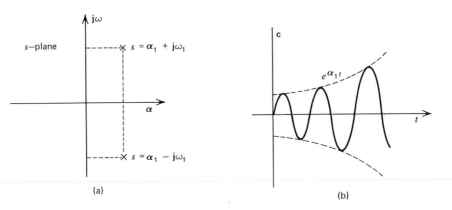

Figure 2.12 (a) Roots of the characteristic equation in the complex plane. (b) Response for a system with positive real components.

2.6 A Control System Example

The application of basic analysis techniques discussed in the last section begins with the control system block diagram. Consider the simplified NC apparatus illustrated schematically in Figure 2.13. An input signal, θ_i, is translated into a voltage by a digital to analog device with transfer function G_1.* This is passed to an amplifier with gain m. The amplifier is used to control a servomotor with transfer function, G_2, producing a torque which is converted to a position displacement, θ_0, by drive component, G_D. A transducer with transfer function, G_f, transmits position feedback.

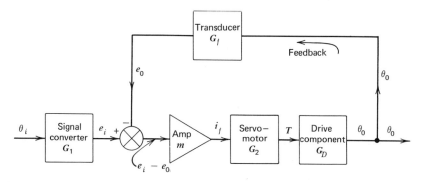

Figure 2.13 Block diagram showing each element of the control system.

Examining the individual transfer functions in more detail,

$$G_1 = \frac{e_i}{\theta_i} \tag{2.12}$$

where e_i is the driving voltage for the amplifier. The gain of the amplifier can be expressed as

$$m = \frac{i_f}{e_i - e_0} \tag{2.13}$$

where i_f is the output current applied to the field of the servomotor. It should be noted that the amplifier is used over its linear range; hence, m is constant.

The servomotor transfer function, G_2, can be expressed as the ratio of output torque, T, and the input current, i_f. Therefore

$$G_2 = T/i_f \tag{2.14}$$

*Generally, the input signal is a *digital* value derived from coded punched tape.

Finally, the transducer transfer function is

$$G_f = \frac{e_0}{\theta_0} \tag{2.15}$$

and the drive component transfer function can be expressed as

$$G_D = \frac{\theta_0}{T} \tag{2.16}$$

Consider a mathematical model of the system. The output torque, T, can be expressed in terms of the product of the angular acceleration and the system inertia, J. Hence,

$$T = \frac{d^2\theta_0}{dt^2} \cdot J$$

or applying Laplace transform notation

$$\mathbf{T} = s^2\boldsymbol{\theta}_0 J$$

and then

$$G_D(s) = \frac{\boldsymbol{\theta}_0}{\mathbf{T}} = \frac{1}{s^2 J} \tag{2.17}$$

Using the rules for evaluating cascaded system elements, the block diagram may be redrawn as shown in Figure 2.14. Thus

$$\frac{mG_2}{Js^2} = \frac{\boldsymbol{\theta}_0}{\mathbf{E}_i - \mathbf{E}_0}$$

and from equations (2.12) and (2.15)

$$\frac{mG_2}{Js^2} = \frac{\boldsymbol{\theta}_0}{G_1\boldsymbol{\theta}_i - G_f\boldsymbol{\theta}_0}$$

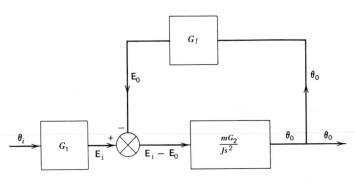

Figure 2.14 Block diagram for transformed system.

For illustrative purposes assume that $G_1 = G_f$ (it should be noted that this is not generally the case*). The above equation can be rewritten

$$\frac{mG_1G_2}{Js^2} = \frac{\theta_0}{\theta_i - \theta_0} \qquad (2.18)$$

Equation (2.18) can be rearranged to yield the transfer function of the control system

$$G(s) = \frac{\theta_0}{\theta_i} = \frac{mG_1G_2/J}{s^2 + mG_1G_2/J} \qquad (2.19)$$

which corresponds in form to equations (2.4) and (2.10).

Assume that the input, θ_i, is a step function such that

$$\theta_i = 0, \qquad \text{for } t \leqslant 0$$
$$\theta_i = A, \qquad \text{for } t > 0$$

From Table 2.1, the Laplace transform for a step function is A/s; therefore, equation (2.19) becomes

$$\theta_0 = \frac{A}{s}\left[\frac{mG_1G_2/J}{s^2 + mG_1G_2/J}\right] \qquad (2.20)$$

Using partial fractions, this becomes

$$\theta_0 = A\left[\frac{1}{s} - \frac{s}{s^2 + (mG_1G_2/J)}\right] \qquad (2.21)$$

Taking the inverse transform of equation (2.21) we get

$$\theta_0 = A(1 - \cos\sqrt{mG_1G_2/J}\,t) \qquad (2.22)$$

It is immediately evident from this equation that the system is not stable. Figure 2.15 shows the response to be oscillatory and not usable in this form. It must be damped so that a quiescent state is attained.

One method of damping the output is to introduce an opposing torque, T_0, proportional to the output shaft speed,

$$T_0 = F\frac{d\theta_0}{dt}$$

or in transformed terms

$$T_0 = sF\theta_0 \qquad (2.23)$$

In the new system

$$\frac{d^2\theta_0}{dt^2} \cdot J = T_{\text{motor}} - T_{\text{damping}} = mG_1G_2(\theta_i - \theta_0) - F\frac{d\theta_0}{dt}$$

---

* This would imply that the input, θ_i, was itself a displacement.

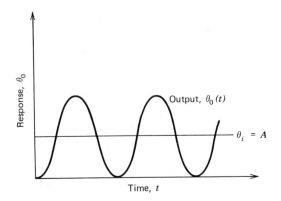

Figure 2.15   Time response for example control system (no damping).

Or, using equation (2.23) and converting to a transformed expression,

$$s^2 J\theta_0 = mG_1 G_2(\theta_i - \theta_0) - sF\theta_0 \tag{2.24}$$

Solving for $G(s)$, it can be shown that

$$G(s) = \frac{\theta_0}{\theta_i} = \frac{1}{Js^2/mG_1G_2 + sF/mG_1G_2 + 1}$$

Letting the undamped natural frequency, ω_n, be expressed using equation (2.22)

$$\omega_n = \sqrt{mG_1G_2/J}$$

and we get

$$G(s) = \frac{\theta_0}{\theta_i} = [s^2/\omega_n^2 + sF/mG_1G_2 + 1]^{-1} \tag{2.25}$$

For the step input, A/s,

$$\theta_0 = \frac{A}{s}[s^2/\omega_n^2 + sF/mG_1G_2 + 1]^{-1}$$

where the characteristic equation is

$$s^2/\omega_n^2 + sF/mG_1G_2 + 1 = 0 \tag{2.26}$$

To simplify the above expression let $\lambda = F\omega_n/2mG_1G_2$. Equation (2.26) is found to have the roots

$$-\lambda\omega_n \pm \omega_n \sqrt{\lambda^2 - 1}$$

As shown in Section 2.5.1 ($\lambda^2 = \beta^2/\alpha^2$), the value of λ^2 will effectively determine the response of the system. Examining values of λ^2:

1. When $\lambda^2 > 1$ (real roots, an overdamped response)

$$\theta_0 = A\left\{1 - \frac{e^{-\lambda\omega_n t}}{\sqrt{\lambda^2 - 1}}\sinh\left(\sqrt{\lambda^2 - 1}\,t + \cosh^{-1}\lambda\right)\right\} \qquad (2.27a)$$

2. When $\lambda < 1$ (imaginary roots, an underdamped response)

$$\theta_0 = A\left\{1 - \frac{e^{-\lambda\omega_n t}}{\sqrt{1 - \lambda^2}}\sin\left(\omega_n\sqrt{1 - \lambda^2}\,t + \cos^{-1}\lambda\right)\right\} \qquad (2.27b)$$

3. When $\lambda = 1$ (equal roots, critically damped response)

$$\theta_0 = A\{1 - (1 + \omega_n t)\,e^{-\omega_n t}\} \qquad (2.27c)$$

Recalling that the value of λ is wholly dependent on control system elements, the motor inertia, and damping force,

$$\lambda = \frac{F}{2\sqrt{J}}[mG_1 G_2]^{-1/2} \qquad (2.28)$$

The response of the system can therefore be controlled by correctly specifying the appropriate element values.

2.6.1 Parameters for System Response

In deciding upon the most suitable system response, two criteria can be applied; the system should respond rapidly to a given input signal, and oscillation about a quiescent state should be minimized.

With critical damping, $\lambda = 1$, oscillation is eliminated but system response is slow (underdamped and overdamped responses satisfy neither criteria). In practical applications, $0.4 \leqslant \lambda \leqslant 0.9$ will result in a slight *overshoot* (see Figure 2.16) but will provide significantly faster system response. The accuracy required by a particular application will dictate the tolerable overshoot.

A damping term, $sF\theta_0$, has been incorporated into the mathematical description of the system, but as yet we have not proposed a method for implementing it in the control loop.

The transformed torque equation (2.24) can be rewritten:

$$s^2 J\theta_0 = mG_1 G_2\theta_i - mG_1 G_2\theta_0 - mG_1 G_2\frac{2\lambda}{\omega_n}s\theta_0$$

or

$$s^2 J\theta_0 = mG_2\left[G_1\theta_i - G_1\theta_0 - G_1\frac{2\lambda}{\omega_n}s\theta_0\right]$$

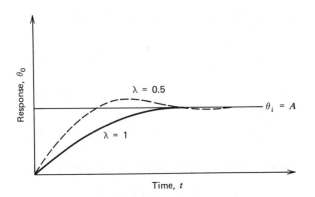

Figure 2.16 Desired system response where λ is the critical damping ratio.

Using equations (2.12) and (2.15) and the idealization $G_f = G_1$, we can rewrite the above expression as

$$s^2 J\theta_0 = mG_2(E_i - E_0) - G_1 \frac{2\lambda}{\omega} s\theta_0 \qquad (2.29)$$

The term $(E_i - E_0)$ is the transformed position error signal transmitted through the feedback loop, as illustrated in Figure 2.14. In equation (2.29) the term $G_1(2\lambda/\omega)s\theta_0$ acts to modify the position error with a velocity dependent signal using a *velocity feedback loop* added to the control system, as shown in Figure 2.17. This velocity feedback is provided by transducer output proportional to velocity, thereby causing the actual damping to be simulated by an electrical sensing device.

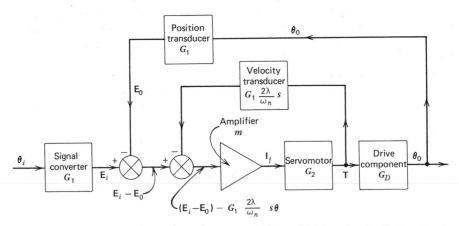

Figure 2.17 Control system block diagram with position and velocity feedback.

References

1. Shinners, S. M., *Modern Control System Theory and Application*, Addison-Wesley, Reading, Mass., 1972.
2. Raven, F. H., *Automatic Control Engineering*, 2nd ed., McGraw-Hill, New York, 1975.
3. Harrison, H. L., and Bollinger, J. G., *Introduction to Automatic Controls*, 2nd ed., International Textbook Co. (Intext), New York, 1969.
4. Kaplan, W., *Elements of Differential Equations*, Addison-Wesley, Reading, Mass., 1964.
5. Ayres, F., Jr., *Differential Equations*, McGraw-Hill (Schaum's Outline Series), New York, 1952.
6. Wylie, C. R., *Advanced Engineering Mathematics*, 4th ed., McGraw-Hill, New York, 1975.
7. Spiegel, M. R., *Laplace Transforms*, McGraw-Hill (Schaum's Outline Series), New York, 1965.

Problems

1. Describe an operation in which a control system could *not* respond as well as a human operator. Develop a general rule governing this situation.
2. Specify the necessary input and output, sensing devices, and drive components required for a control system to maintain constant highway speed in an automobile. Illustrate your system with a block diagram showing the input and output of each element in the control loop.
3. Would an open loop control system be suitable for a conveyor system that always has the same loading per unit length and must maintain constant speed? Explain your answer.
4. A particular positioning control system is designed to move a probe 0.025 mm in 1 msec. Input signals in the form of pulses indicate movement, and 1 pulse corresponds to 0.025 mm of movement. Assuming all motion to be in the positive direction, plot motion with respect to time when the machine receives: (a) 500 pulses/sec, (b) 1000 pulses/sec, and (c) 2000 pulses/sec. What is the time constant, τ, for this system?
5. The response of a system is shown to be

    $$y = A(1 - e^{-\alpha t})$$

    (a) What is the value of τ for this system?
    (b) How could the characteristic response time of the system be increased? Assume that the characteristic response is measured as the time required for $y = 0.99A$.
6. Describe three types of mechanical amplifiers and indicate an expression for gain in each.

7. Consider the electrical circuit illustrated in Figure P.2.1. The gain of the amplifier is m and the transfer function of the resistor is Ω. If the input voltage, E_{in}, and the feedback voltage, E_{out}, are in phase, how does the output of the amplifier behave? What is a common consequence of this amplifier behavior?

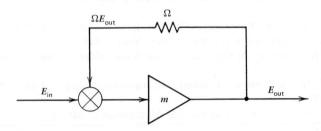

Figure P.2.1 Problem 2.7.

8. Show that equation (2.3) in the text is true for any input or output condition.
9. Define the system transfer function for the closed loop control shown in Figure P.2.2. How would the transfer function change if a second feedback loop with transfer function Φ_5 were placed between points A and B?

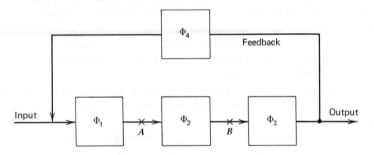

Figure P.2.2 Problem 2.9.

10. Determine the transfer function for the R-R-C circuit illustrated in Figure P.2.3. How would an increase in R_2 affect the value of the transfer function?
11. What is the transfer function for a simple spring-mass system where a mass m is attached to a spring (constant $= k$)? What is the significance of the transfer function in this system?
12. Using the Laplace transform prove that:

$$\text{(a)} \quad \mathscr{L}\{t \cos \omega t\} = \frac{s^2 - \omega^2}{(s^2 + \omega^2)^2}$$

$$\text{(b)} \quad \mathscr{L}\{t^2 \sin t\} = \frac{6s^2}{(s^2 + 1)^3}$$

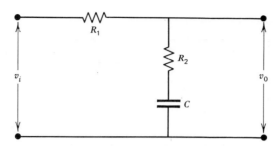

Figure P.2.3 Problem 2.10.

13. What is the Laplace transform of the R-R-C circuit in Problem 10?

14. Use Laplace transform methods to evaluate the displacement and velocity of the spring-mass system described in Problem 11. Let the initial displacement $x(0) = 5$ and $x'(0) = 2$. How could the oscillatory response of this system be shifted to a higher frequency?

15. Find

$$\mathcal{L}^{-1}\left\{\frac{5s+4}{s^2}+\frac{2s-18}{s^2+9}+\frac{8-6s}{16s^2+9}\right\}.$$

What type of response is represented by this transform?

16. What kind of response can be expected from a system described by

$$\frac{d^2x}{dt^2}+2\frac{dx}{dt}+4x=0$$

with $x(0) = x_0$, $x'(0) = 0$? How would the above expression look for a critically damped response?

17. An open loop system has two elements with transfer functions

$$G_1(s) = s+1;\ G_2(s) = \frac{1}{s^2+s+1}$$

respectively. Evaluate the open loop transfer function and the system response as a function of time. Assume a step function A as input at $t = 0^+$.

18. A closed loop control system has three elements as illustrated in Figure P.2.4. What is (a) the transfer function of this system? (b) the output response as a function of time?

19. A closed loop control system is illustrated in Figure P.2.5. Write the transfer function for this system. If all the feedback elements had transfer functions $F_1 = F_2 = F_3 = 1$, how would the system transfer function change?

20. A newly designed control system element is found to have the transfer function $[s^2 + (\beta - \alpha)s - \alpha\beta]^{-1}$ where α and β are positive quantities. Is the component stable?

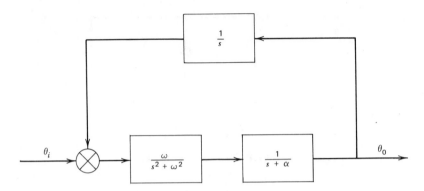

Figure P.2.4 Problem 2.18.

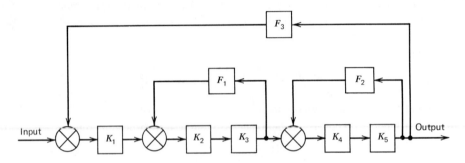

Figure P.2.5 Problem 2.19.

21. A control system is modeled by the equation

$$\frac{d^2y}{dt^2} + \alpha y = \beta \frac{dy}{dt} + A \cos \omega t$$

What is the transfer function for this system when $y(0) = Y_0$, $y'(0) = 0$? Discuss how the values α, β, and A affect the system stability.

22. Calculate the response of the system described in Problem 21 if $\alpha = 4.6$, $\beta = 3.0$, and $A = 6$.

23. A system is described by the equation

$$\frac{1}{4}\frac{d^2x}{dt^2} + \frac{9}{4}x = \delta(t)$$

where $\delta(t)$ is the unit impulse at $t = 0$. What is the time response of this system? Is it stable? *Note*: $\mathcal{L}^{-1}\delta(t) = 1$.

24. A flywheel is positioned by a control system. The motor applies a torque to the flywheel of 225 N · m/rad of misalignment, and a flywheel velocity of 1 rad/sec produces a damping torque of 50 N · m. If the system moment of inertia is 0.3 kg · m^2 and the input position is suddenly rotated $\pi/2$ rad at $t = 0^+$, find an expression for the position, θ, of the flywheel at any time $t > 0$.

25. Show that any second order positioning control system that lacks a damping term (i.e., velocity feedback) will be unstable.

26. Develop a differential equation that is a model for the Watt speed governor illustrated in Figure 2.3. Use Laplace transforms to predict the motion of the sleeve for a step increase in motor shaft speed of Ω_0 at time $t = 0^+$.

27. In the example control system of Section 2.6 the simplification $G_1 = G_f$ was made. Analyze the system for $G_1 \neq G_f$ in a manner analogous to equations (2.18) through (2.29).

28. Draw the block diagram of a control system which enables a moving vehicle to follow a given path by sensing the location of a wire buried in the roadway. Discuss the following aspects: (a) types of sensors, (b) stability requirements, and (c) practicality.

Chapter Three
NC Machine
Control Systems

The control section of a servomechanism which uses discrete input signals and digital logic circuits to cause the system to perform is termed a *numerical control*. Numerical controls are identical conceptually to the automatic feedback control systems discussed in Chapter Two with one fundamental difference in operation. The controls discussed earlier functioned depending upon the magnitude or amount of the input signal. Such devices are generally termed *analog* controls. A numerical control uses only the presence or absence of signals to operate. Hence, it is a *digital* device, which accepts coded input data that can be translated into one of two opposing states (e.g., *on* or *off*).

It is important that the differences between analog and digital input are understood. Consider a device that is used to position a slide. Referring to Figure 3.1*a*, a potentiometer is used to vary the voltage to the control system servomotor. As long as a functional relationship between slide position and voltage can be developed, the potentiometer can be calibrated in terms of distance, and the resistance contact can be moved to provide the desired input signal.

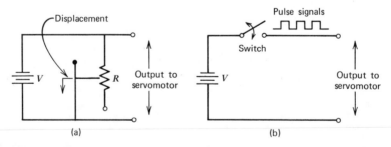

Figure 3.1 Analog and digital signals. Output voltage for analog is a function of displacement; output voltage for digital is a pulse function. (*a*) Analog signal. (*b*) Digital signal.

The digital approach to this positioning problem is illustrated in Figure 3.1*b*. To generate the desired input, a switching device is used. Each time the switch is closed, an electrical pulse is sent to a servomotor which moves the slide a constant amount. The switch is sequentially opened and closed, thereby generating *discrete pulses* until the desired position is attained.

The NC system accepts coded data in the form of punched or magnetic tape, or in the case of DNC or CNC*, binary data sent over computer communications lines. Each block of data is stored and then transformed into the appropriate pulse signals. For a positioning command, the pulses are used to control a servomotor which powers the required drive components. Because input is in discrete pulse form, feedback must be similarly configured. The input signal is therefore modified by transducer feedback which translates position and/or velocity in an equivalent pulse stream. An NC control loop is schematically illustrated in Figure 3.2.

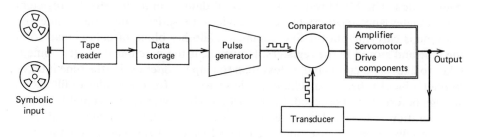

Figure 3.2 Schematic of digital information flow for an NC loop.

3.1 Elements of the NC System

3.1.1 The Machine Control Unit

The machine control unit (MCU) is a special purpose computer. The functions that it performs vary with the complexity of the machine tool and the tasks that it is required to execute. Every machine control performs those functions attributed to an automatic control system.

The MCU decodes input data and stores it in a *memory*. It *communicates* with the control loop servo components and *senses* data returned from feedback elements. Most important, it executes commands based on *logical decisions* derived from all system information.

* Direct Numerical Control (DNC) and Computer Numerical Control (CNC) are discussed in detail in Chapter Ten.

The sophistication of the MCU determines the operations that can be performed by the NC machine tool. The MCU can be programmed* to handle a wide variety of tasks. For example, it can be responsible for the control of lubricants during metal cutting—a simple *on* or *off* command. It can also make the complex calculations required for acceleration/deceleration control and linear/circular or parabolic interpolation. Because punched tape is used in a majority of NC applications, a typical MCU receives its input data from a tape reading device. One such device is the photoelectric tape reader that can interpret one hundred to one thousand punched characters per second from a perforated tape passed under a *reading head*. The reading head consists of a light source and an array of light sensitive elements. As a hole passes over the light source, an element of the array is turned *on*. The pattern of *on* elements is decoded by the MCU and is either acted upon immediately or transferred to the information storage section of the MCU.

The NC machine responds to instructions which are held in storage registers until needed. The MCU retrieves a block of data from *active* storage registers and actuates the appropriate servo devices to satisfy the data. While the machine is carrying out the first command, another block of data is read from the tape and stored in the registers. However, if the *machine cycle* (the time it takes to execute a command) is less than the tape *reader cycle* (the time it takes to read a block) the NC machine will have to wait for the reader to fill active storage before it may proceed. Such start-stop behavior is unacceptable. When the operation of the machine is retarded, the surface that is machined will show *dwell* marks at each pause, and the control system itself cannot operate in a continuous fashion. To eliminate this start-stop behavior, *buffer storage* has been developed.

When buffer storage is used, decoded data from the tape reader is read into a series of buffer registers. These registers serve as a temporary storage location for input data. As the NC machine executes a block of data, the next block is stored in the buffer registers and is electronically switched to the active registers as soon as the MCU has cleared the latter. In this way data input and command execution occur simultaneously and start-stop behavior is eliminated.

To generate an actuation signal, the MCU must convert the data stored in the active registers to a pulsing signal which represents that data. The number of pulses and the rate at which they are generated are functions of the distance to be traveled and the cutter feedrate. The distance to which a single pulse corresponds is the smallest dimensional resolution of the NC machine. In other words, if a single pulse represents a distance of 0.002 mm, commands to the

---

* Most MCUs are *hardwire* programmed; that is, circuitry within the unit is set up to perform specific tasks. However, current emphasis is on *softwired* MCUs (Chapter Ten).

servomotor would be in increments of 0.002 mm. Since the MCU cannot generate fractional pulses, a movement of less than 0.002 mm could not be attained.

If an active register contains the representation of an incremental movement of 50 mm along the x-axis, the MCU would generate 25,000 pulses (1 pulse = 0.002 mm) and, if a feedrate of 1 meter per minute (mpm) is required, the pulses would be metered to span 0.05 minute (approximately 8333 pulses/sec). Pulse generation is complicated by vector motion in two or more axes. The pulses controlling each axis of motion must be timed so that the proper vector direction and the desired feedrate are maintained. For example, if the x component is three times the y component for a given 2-D vector, the number of x-axis pulses would be three times the number of y-axis pulses over the same time span.

3.1.2 Feedback in the NC Loop

The availability of feedback in an NC system is a function of the accuracy that the system must maintain and the loading on the moving elements. From Chapter Two we should recall that *open loop* controls are characterized by the absence of feedback, whereas *closed loop* controls use feedback to modify the command signal. To illustrate open and closed loop control systems for NC machines, consider a positioning machine that moves a table to exact coordinate locations.

Figure 3.3a illustrates an open loop positioning system. The command signal is converted into an appropriate number of pulses by the MCU. As each pulse is transmitted, a subtraction circuit is used to reduce the command count by one. When the command count reaches zero, the appropriate number of pulses has been sent and the table is *assumed* to be in the correct location.

In a closed loop positioning system, illustrated in Figure 3.3b, the MCU generates a pulse signal until the signal returned from the feedback transducer agrees with the original number of pulses required to execute the movement. A comparator is used to compare the count of feedback pulses with the original value, and an error signal is output until the feedback pulse count agrees with the input value.

The above example shows that if machine drive components deteriorate due to wear or other reasons, the open loop NC system would not recognize any difference in table position since it counts pulses. For applications in which loading does not vary and in which the frictional and wear characteristics of the servo drives are well known, open loop NC can provide good results. Such systems are generally much less costly than comparable closed loop machines.

Closed loop NC systems are used when a high degree of accuracy is required, regardless of machine loading or wear. Closed loop systems use *direct*

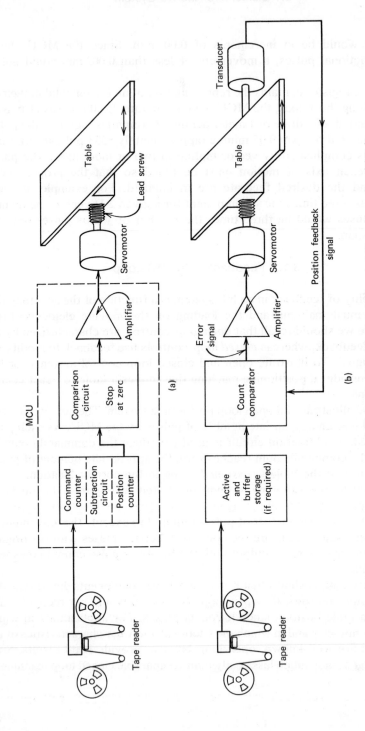

Figure 3.3 Open and closed loop positioning systems. (*a*) Open loop positioning control. (*b*) Closed loop positioning control.

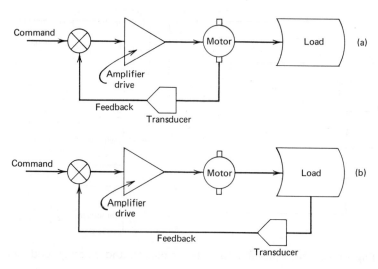

Figure 3.4 Schematic diagram of (a) indirect feedback control, and (b) direct feedback control.

and *indirect* feedback. The indirect feedback method measures the output of the servomotor, as illustrated in Figure 3.4a. Although it is a common closed loop configuration, indirect feedback does not provide as high a degree of accuracy as the direct feedback approach, which also includes the load in the feedback loop. In a closed loop with direct feedback (Figure 3.4b), transducers are used to monitor the load (e.g., the table in Figure 3.3b). Direct feedback is more accurate and more costly to implement.

3.1.3 Position and Velocity Feedback

Most closed loop servomechanisms include both position and velocity control. Position feedback in an NC system is in the form of pulses that are compared to actual command data. The result of this comparison is the error signal. Velocity control is essential in NC applications in which it is important to maintain high levels of accuracy for final position, surface finish, and path accuracy.

Figure 3.5 shows the basic elements of a closed loop NC system with both position and velocity feedback. Pulse input is sent to the comparison counter for position feedback, and the error signal is output. In the system illustrated, the servos are driven by analog voltage; hence, the digital to analog device converts the pulsed error signal to an analog voltage which is passed to a control amplifier for further processing. It should be noted that the *compensa-*

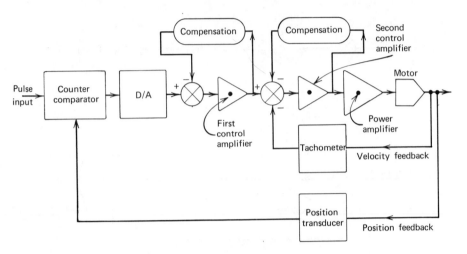

Figure 3.5 NC system schematic—position and velocity feedback.

tion loops are a form of internal feedback that improves the performance of the control system.*

The output of the first control amplifier is modified by feedback from a velocity transducer. Velocity feedback is generally provided by an ac or dc tachometer coupled to the servomotor. Unlike the position transducer which outputs a digital pulse signal, the tachometer generates an analog voltage that is compared with the analog signal of the first control amplifier. Coupled with the second control amplifier, the velocity loop feedback essentially magnifies the gain for the input to the power amplifier.

The power amplifier provides input to the servomotor. It may be either electromechanical, pneumatic, or hydraulic, depending upon the drive system being used. Motors and power amplification systems are discussed in Chapter Four.

It should be noted that a position feedback loop provides some velocity control even without a velocity feedback loop. If the position feedback signal lags behind the command signal, an increasingly large error signal will be produced. This larger error signal will cause the motor to operate at a higher rotational velocity. However, information must be processed through the longer position feedback loop, causing velocity control response to be relatively slow. The shorter velocity feedback loop provides immediate correction for velocity changes, and maintains a stabilizing effect on the servo system.

---

* For further information on compensation and internal feedback, the reader is referred to Reference 1.

3.1.4 Sensitivity

An NC system, as illustrated in Figure 3.5, is made up of individual elements with individual transfer functions. If the transfer function of each element is known, we can use the methods outlined in Chapter Two to develop the system transfer functions, $G(s)$. In the analysis of a control loop, it is important to predict the effect of changes in a given element on the entire system. The environment in which NC equipment operates can cause changes in system elements due to temperature, friction, or wear, and the magnitude and effect of these changes should be known in advance.

Sensitivity is the measure of the relationship between the entire system's characteristics and the characteristic of a single element. Letting G be the system transfer function and H be the transfer function of a single element, the *differential sensitivity* is defined as:

$$S_{G,H} = \frac{dG/G}{dH/H} \tag{3.1}$$

Equation (3.1) is a ratio of the percent change in the system transfer function to the percent change in the element transfer function. It is important to note that the above equation is valid only for small changes, and an ideal closed loop system would have zero sensitivity to any element.

To examine the concept of sensitivity in greater detail, let us consider the abbreviated block diagram of an NC positioning system illustrated in Figure 3.6. Referring also to Figure 3.5, H_1 represents the transfer function of the input elements (the count comparator and D/A converter); H_2 represents the characteristic of the position transducer in the feedback loop; and J represents the combined transfer function of the control amplifiers, compensation network, power amplifier, and motor. To simplify the analysis the velocity control loop is considered part of J.

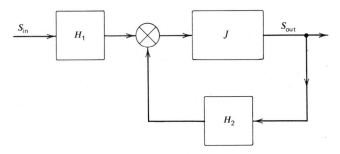

Figure 3.6 A simplified transfer function representation of system in Figure 3.5.

We have shown that the overall system transfer function G is given by

$$G = \frac{S_{out}}{S_{in}} = \frac{H_1 J}{1 + H_2 J} \qquad (3.2)$$

Using equations (3.1) and (3.2) we can now consider the sensitivity of the system transfer function to H_1, H_2, and J.

To determine the sensitivity of G with respect to J we write

$$S_{GJ} = \frac{dG/G}{dJ/J} = \frac{J}{G} \frac{dG}{dJ}$$

where

$$\frac{dG}{dJ} = \frac{(1 + H_2 J)H_1 - H_1 J H_2}{(1 + H_2 J)^2} = \frac{H_1}{(1 + H_2 J)^2}$$

or

$$S_{GJ} = \frac{J}{G} \frac{dG}{dJ} = \frac{1}{1 + H_2 J} \qquad (3.3)$$

Equation (3.3) shows that the sensitivity of the entire system to the combined transfer function J is inversely proportional to $1 + H_2 J$. From a sensitivity standpoint, S_{GJ} should be much less than 1. Therefore, a large value for J would be highly desirable.

To determine the sensitivity of G with respect to H_1,

$$S_{G,H_1} = \frac{H_1}{G} \frac{dG}{dH_1}$$

where

$$\frac{dG}{dH_1} = \frac{J}{1 + H_2 J} = \frac{G}{H_1}$$

Therefore

$$S_{G,H_1} = \frac{H_1}{G} \frac{dG}{dH_1} = 1 \qquad (3.4)$$

Similarly, to determine the sensitivity of G with respect to H_2

$$S_{G,H_2} = \frac{H_2}{G} \frac{dG}{dH_2}$$

where

$$\frac{dG}{dH_2} = -\frac{H_1 J^2}{(1 + H_2 J)^2} = \frac{-H_1^2 J^2}{H_1(1 + H_2 J)^2}$$

Therefore

$$S_{G,H_2} = \frac{H_2}{G} \cdot \frac{-H_1^2 J^2}{H_1(1 + H_2 J)}$$

Upon rearranging

$$S_{G,H_2} = \frac{-H_2 J}{1 + H_2 J}$$ (3.5)

For cases in which $H_2 J \gg 1$, $S_{G,H_2} \approx -1$.

Equations (3.4) and (3.5) show that the characteristics of the input and feedback transducers are critical. Any change in either H_1 or H_2 will be directly reflected in a change in overall system response. Therefore, the elements must possess stable characteristics regardless of operating environment. In the next section (and again in detail in Chapter Four) we examine some of the elements that are represented by H_1 and H_2.

3.2 Feedback Components—A General Description

3.2.1 Transducers

A transducer can be defined as any sensing device which translates a measured quantity into some other measurable quantity. The NC system uses both velocity (analog) and position (digital) transducers. An analog transducer produces a signal that varies in direct proportion to the measured quantity, whereas a digital transducer generates a pulsed (*on* or *off*) signal which represents a discrete portion of the quantity to be measured.*

In NC applications a transducer is selected based on the following criteria:

1. The type of compatibility required by other components in the loop (e.g., is digital or analog output required?).
2. The quantity to be measured.
3. The environment in which the measurement must take place.

To be most effective, a transducer should make its measurements directly without gearing or linkages, as such appendages introduce possible mechanical errors.

Sensing devices in the position feedback loop are either *linear* or *rotary* transducers. Because position feedback is digital, these transducers must translate linear or rotary motion into a pulsed output signal.

All digital transducers, regardless of type, are pulse generators. Consider the simple linear transducer illustrated in Figure 3.7a containing a coded scale with very fine gradations attached to the table. An optical or magnetic reading head emits a pulse as it senses each line. The individual pulses are summed to indicate the total displacement of the table.

* A third type of transducer, called the numerical encoder, is also available (see Chapter Four).

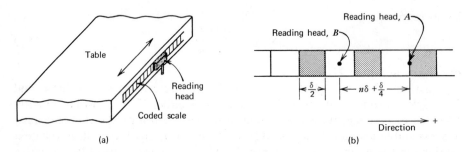

Figure 3.7 (a) A linear digital transducer. (b) Method for direction discrimination used in linear digital transducers.

Because the form of the pulses is independent of the direction of translation, a *direction discriminator* is used to allow a counter to add or subtract pulses, depending on the direction of translation. As seen in Figure 3.7b, if gratings are a distance $\delta/2$ apart, two reading heads are placed at a distance $n\delta + \delta/4$ apart. If the displacement is in the direction shown, where a shaded area represents a **1** pulse and the light areas represent a **0** pulse, each reading head will change state (i.e., go from **0** to **1** or **1** to **0**) in a unique manner illustrated by Table 3.1. Hence, when a change of state is sensed, the direction of motion may be easily determined based on the information in the table (Reference 2).

Table 3.1
Direction Discriminator Values

	A^a changes state		B^a changes state	
	0 to 1	1 to 0	0 to 1	1 to 0
A reads 0			+	−
A reads 1			−	+
B reads 0	−	+		
B reads 1	+	−		

[a]See Figure 3.7b.

The rotary digital transducer operates on the same basic principle as the linear transducer. However, input to this transducer may be taken directly from the drive leadscrew or through a geared rack and pinion arrangement attached to the table.

3.2.2 Power Amplifiers and Actuators

Amplification takes two distinct forms on numerically controlled equipment. The input signal is increased by one or more *high impedance voltage amplifiers* to provide the necessary voltage to drive the power amplifier. It is the low impedance power amplifier that supplies the necessary input energy to the servomotor.

Power amplification may be electromechanical, hydraulic, or pneumatic. The choice of systems depends on the motor horsepower to size ratio, the motor speed, and temperature considerations. To illustrate a typical power amplifier arrangement, consider the hydraulic system represented schematically in Figure 3.8.

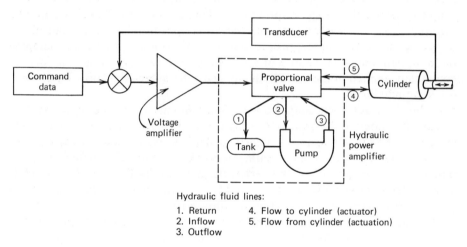

Hydraulic fluid lines:

1. Return   4. Flow to cylinder (actuator)
2. Inflow   5. Flow from cylinder (actuation)
3. Outflow

Figure 3.8 Hydraulic actuator. Electrical signal is converted to fluid "signal" with hydraulic power amplifier.

An error signal is passed through the voltage amplifier to an electronically controlled valve whose output is directly proportional to the input voltage signal. The valve opens to allow a measured volume of hydraulic fluid under high pressure to drive the cylinder which in turn moves the table. Power amplification is accomplished by the opening or closing of the valve. Hence, in this system the valve is a power amplifier.

Actuation components include a broad range of devices: electric, hydraulic, and pneumatic motors as well as cylinders. The choice of power actuator is matched to the type of power amplification system which is used.

3.2.3 Error Signal Recognition

In NC systems error signals enter the forward flow of information at the position and velocity feedback loops. Referring again to Figure 3.5, it should be noted that the gain of each control amplifier plays an important role in the overall response of the system to the error signal.

Consider the velocity feedback loop. An error signal results when the tachometer output does not equal the input velocity command. The second control amplifier (Figure 3.5) serves to amplify the effect of the error signal. Depending on the gain of the amplifier, a very small error signal can produce large changes in signal to the power amplifier, causing subsequently large velocity fluctuations.

The gain of the first control amplifier (Figure 3.5) must be such that the servomechanism does not *oscillate* or *hunt* about the command position. In a simple mechanical analogy, a spring-mass damper system will oscillate about a zero point (the quiescent point) until the *at rest* position has been attained. As we saw in the last chapter, proper selection of spring stiffness and damping factor determines the response of the system. If, however, an outside force is continually applied to the system, damping can be overcome, and the mass will oscillate indefinitely.

In the NC positioning loop the amplifier gain can be viewed as the system forcing function. If the gain is too high, the intrinsic damping effect of the system will be overcome and sustained oscillation results. In practice, the error signal is modified by the compensating networks (Figure 3.5). By increasing the time constant (response) of the servo, these networks limit oscillation while allowing the relatively high gain required for positioning accuracy. The result is a slower response when correcting position errors.

The positioning loop is also affected by a mechanical characteristic, called *backlash*, in the drive components and transducer gearing. All mechanical drive components move a small amount before the individual elements make contact and force is generated.* When backlash is severe, the system encounters no force resistance as the initial motion command is encountered and the servomotor can then build up enough speed to cause overshoot. The system will attempt to correct the overshoot by reversing this sequence, but backlash again causes inaccuracy. When friction is low, resulting in low force resistance, the servo will oscillate about the command location.

Although mechanical backlash cannot be eliminated, it can be reduced. All numerical control servos exhibit a *dead-zone* which prohibits very small

---

*As an analogy, consider an automobile steering wheel. In most standard models, the wheel can be turned a fraction of a revolution before the car responds—this is backlash in the steering mechanism.

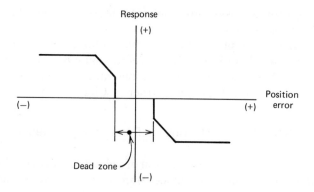

Figure 3.9 Dead-zone for servo response.

position errors from activating the servomotor. A dead-zone is represented graphically as shown in Figure 3.9.

The hydraulic power amplification system discussed earlier, when its valve is completely closed, has an inherent dead-zone; that is, very small signals from the power amplifier will cause the valve to move, but not enough to allow any hydraulic fluid to reach the cylinder.

In many NC systems the dead-zone response is adjustable. For example, a needle valve may be connected between the pressure and return lines for hydraulic fluid. The needle valve allows a small amount of leakage to be overcome before pressure is supplied to the cylinder. The magnitude of this leakage can be set by adjusting the needle valve, thereby determining the width of the dead-zone.

3.3 Positioning Control Systems

Positioning control systems have definite restrictions on the vector paths which the cutting tool may be commanded to follow. In the simplest positioning systems, the path which the cutter follows between points is not *programma-ble*. Thus, only the point locations may be specified. Positioning systems that move from some starting point to a final position without path control are called *point-to-point* (PTP) machines. NC systems with PTP controls generally perform operations which are *path independent* (e.g., drilling or spot welding).

The machine control unit in a PTP system contains registers which hold the individual axis motion commands. In some systems, the x-axis command is satisfied initially, followed by the y- and z-axis commands. This operation may produce a zigzag path which will ultimately terminate at the proper point location.

Many NC positioning systems contain a more complex MCU. In these servos, positioning commands are evaluated simultaneously so that vector motion in two axes is possible. However, this vector motion is limited to a one-to-one pulse output. Therefore, only 45° vectors may be traced. Such systems are sometimes called *straight cut*. The operations available for PTP and straight-cut machines are discussed further in Chapter Seven.

3.3.1 Velocity Control in Positioning Systems

Most positioning systems have some form of velocity control to eliminate *over-* or *undershoot* errors and to reduce transit time between points. When point-to-point NC equipment is used, the traverse time between points may be rapid (3 to 8 mpm) to reduce total manufacturing time. Because PTP operations occur only at the final position, significant position overshoot may be acceptable, provided subsequent commands based on feedback cause the cutter to rapidly converge on the proper position. The position response of such systems is illustrated in Figure 3.10a. Once a position error of magnitude ϵ has been reached, the system is within the feedback loop dead-zone and further response cannot occur.

The effects of backlash can be controlled by proper velocity control. To eliminate the effect of backlash, the tool traverses at velocity V_1 to some overshoot position, A (Figure 3.10b). At this point one or two oscillations at lower intermediate velocity V_2 are carried out so that the tool always approaches the command position from the same direction. This eliminates the need for direction reversal (involving backlash) when the error magnitude is approaching the width of the dead-zone.

For positioning systems in which overshoot must be minimized, *velocity step* control may be implemented. The initial velocity, V_1, is decreased at a certain command position. As the tool moves closer to the final location, the velocity further decreases in steps. The position response of such a system is shown in Figure 3.10c (Reference 2).

3.3.2 Position Error

Earlier in this chapter, the relationship between control amplifier gain and error signal recognition was discussed. Amplifier gain also plays an important role in the determination of the position error, ϵ, which remains once the system has reached a quiescent state.

Defining ϵ as the position error remaining, then

$$\epsilon \propto \frac{\text{mechanical friction of drive components}}{\text{position loop gain}} \quad (3.6)$$

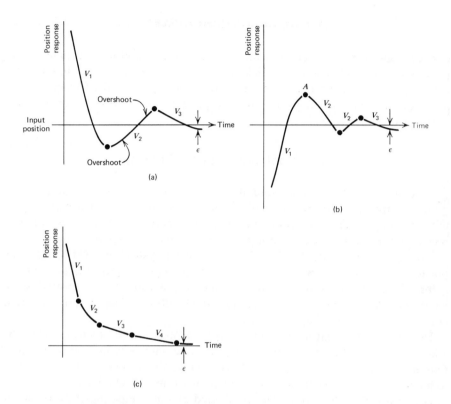

Figure 3.10 Velocity control methods for positioning NC. (*a*) Underdamped response. (*b*) Underdamped response corrected for backlash. (*c*) Step response.

Since it is impossible to make the mechanical friction zero or the position loop gain infinite, a small positioning error will always be present. To reduce this, machine tool builders attempt to reduce friction in their equipment, whereas control systems are designed with high position loop gain. The overall design is a compromise, since high gain can lead to system instability, and the bearings and lubrication devices required to reduce friction increase the cost of the equipment.

The velocity control loop also plays an indirect role in determining the position error, since the velocity control loop increases the gain of the position loop and increases the response time of the system. Faster response time means that the mechanical time constants in the position loop may be decreased without loss of system stability. Hence, from expression (3.6), the position loop gain will increase with a proportional decrease in position error, ϵ.

3.4 Contouring Control Systems

The contouring facility enables an NC machine to follow any path at any prescribed feedrate. The contouring control system manages the simultaneous motion of the cutting tool in two, three, four, or five axes (the fourth and fifth axes are angular orientations) by interpolating the proper path between prescribed points.

3.4.1 Contouring System Elements

Computations performed continuously by the machine control unit are required to precisely control the path of the cutting tool so that it may attain a minimum of *path error*. In a contouring system, therefore, the control system is concerned not only with positioning error but also with path error.

The significant elements of an NC contouring system are: (1) MCU interpolator, (2) comparators, and (3) servo components. Another important element of the total system is the computer software which provides the instructions to drive the contouring system. Details of this aspect of numerical control are given in Chapter Seven.

The MCU interpolator generally consists of hardwired electronic circuitry which performs linear, circular, or parabolic interpolation between input points. Not all contouring systems have each of the above facilities, but, as a minimum, linear interpolation between points must be available.

The feedback signals which are processed by the comparator take on added significance in an NC contouring system. When the mechanical system *lags* behind the command signal, path error will occur unless acceleration and deceleration are precisely controlled.

The servo components of a contouring system are in many ways similar to those found in positioning systems. Because the control of the system is more complicated and therefore more difficult, care must be taken to insure that individual elements are stable over a broad range of operating conditions.

3.4.2 Interpolation

The two most common types of interpolators found in practice are those providing linear and circular facilities. Parabolic and cubic units are also available for use in the more advanced systems.

Linear Interpolation. The term *linear interpolation* may be defined as a method that develops intermediate coordinate points on a straight line between given start and finish coordinates. The input media to an NC machine contain discrete information in the form of absolute coordinates or incremental movements. Because the NC machine is constrained to move along an axial direction in increments no smaller than the distance corresponding to a single pulse, the term *quasi-continuous movement* is applied to indicate that a straight line is approximated by a series of small steps which are coordinated by the interpolator and the MCU.

For example, consider an NC machine in which 1 pulse equals 0.002 mm. For a linear movement between two points shown in Figure 3.11, the machine must move 10 pulses in the X-direction while simultaneously moving 20 pulses in the Y-direction. The intermediate points are determined by the linear interpolator. The output motion would not be the stair-step motion illustrated in the figure because the simultaneous operation of servomotors for both axes will cause a smoother curve. When one considers that the maximum normal displacement from any exact line is less than 0.002 mm, it is evident that sufficient accuracy is attained.

The ability to continuously control the linear movement between points requires the components of velocity that yield motion to be defined. For

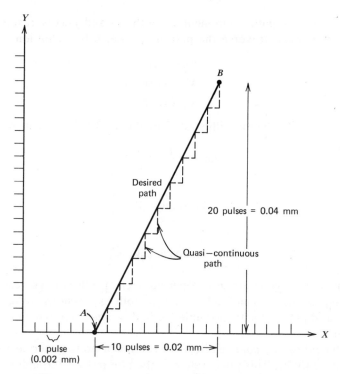

Figure 3.11 Path generated during linear interpolation.

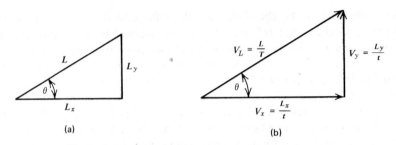

Figure 3.12 (a) Displacement and (b) velocity components for linear interpolation.

simplicity a two-dimensional example is used; however, an extension of the relations to three dimensions is not difficult.

Given a vector motion along a path of length, L, as illustrated in Figure 3.12a, it is seen that

$$\sin \theta = \frac{L_y}{L} \quad \text{and} \quad \cos \theta = \frac{L_x}{L}$$

where L_x, L_y are the path component along the x- and y-axes, respectively. If an NC machine must traverse the path in a time, t, by definition

$$V_L = L/t$$
$$V_x = L_x/t$$
$$V_y = L_y/t$$

From the velocity diagram illustrated in Figure 3.12b, we can write

$$\sin \theta = \frac{L_y}{L} = \frac{V_y}{V_L}$$

or

$$V_y = V_L \sin \theta = V_L \cdot \frac{L_y}{L} \qquad (3.7)$$

Likewise

$$V_x = V_L \cos \theta = V_L \cdot \frac{L_x}{L} \qquad (3.8)$$

The control must generate the proper number of pulses to correspond to V_x and V_y in equations (3.7) and (3.8). Because both expressions contain the term V_L, the desired path velocity, a simple method can be used to obtain V_x and V_y (Reference 3).

A *variable pulse rate control circuit* (VPRC) is used to generate a pulse rate corresponding to V_L. This pulse rate can then be passed through two circuits which effectively multiply the pulse rate by $\sin \theta$ and $\cos \theta$. The output of this

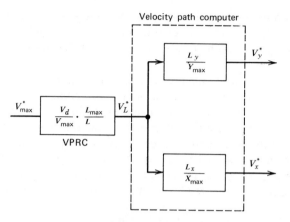

Figure 3.13 Pulse rate generation for linear path velocity. (From Ertell, G., *Numerical Control*, copyright © 1969 by John Wiley & Sons. Reproduced with permission.)

path velocity computer is the required pulse rate for motion along the x- and y-axes.

Referring to Figure 3.13, V^*_{max} is the input to the VPRC corresponding to a pulse rate which produces maximum feedrate for the NC machine. [Note: all starred (*) quantities in this section are pulse rates.] The desired path velocity is input and is stored in a buffer such that the output of the VPRC is V^*_L.

To eliminate the computation of θ, expressions can be developed that are a function of machine maximum velocity, displacements, and the desired feedrate, V_c. If the maximum displacements are defined such that

$$X_{max} = Y_{max} = L_{max}$$

then equations (3.7) and (3.8) can be rewritten as

$$V^*_x = V^*_{max} \cdot \frac{V_c L_{max}/L}{V_{max}} \cdot \frac{L_x}{X_{max}} \qquad (3.9)$$

Likewise

$$V^*_y = V^*_{max} \cdot \frac{V_c L_{max}/L}{V_{max}} \cdot \frac{L_y}{Y_{max}} \qquad (3.10)$$

Using these equations, the velocity path computer is defined. Because the L_x and L_y values must be computed to preset the distance counters, the use of the circuits represented by Figure 3.13 requires no new information.

Circular Interpolation. Because a circular arc can be approximated by a many sided polygon, a linear interpolation system can accurately follow a circular path by moving along line segments (chords) which approximate the arc. The start and end points for each line segment defining the circle are computed, and since each line segment comprises one block of NC data, a large number of blocks must be read at high speeds to assure that the desired path velocity is maintained.

Given the end points of an arc and the arc radius, circular interpolation systems define a circular arc. A *circular path computer* within the MCU eliminates the need to define an arc using individually specified line segment blocks. Not only does a circular interpolation capability simplify the specification of an arc path but it also assures required path velocity without dwell.

Consider the circular arc with appropriate path velocities illustrated in Figure 3.14. From the geometry, expressions for the x and y velocity components of the tangential arc velocity (feedrate), V_c, can be calculated as follows:

$$V_x = V_c \sin \alpha$$
$$V_y = V_c \cos \alpha \tag{3.11}$$

and from the geometry

$$\cos \alpha = I/R = V_y/V_c \tag{3.12a}$$

$$\sin \alpha = J/R = V_x/V_c \tag{3.12b}$$

where I and J are incremented displacements from the arc center to the current point on the arc.

The motion along a circular arc path can be generated in a manner similar to the method used to define straight line segments. A VPRC is used to generate a

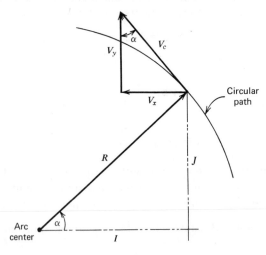

Figure 3.14 Path nomenclature for circular interpolation.

pulse rate corresponding to V_c. This pulse rate is then passed through two circuits which effectively multiply the pulse rate by $\cos \alpha$ and $\sin \alpha$ to obtain V_y and V_x. Unlike the linear path computer in which the angle θ is constant, the circular path computer must vary angle α (or equivalent distance ratios) uniformly with time.

Before we discuss how this uniform variance is accomplished, let us eliminate α from equations (3.11). From equations (3.11) and (3.12)

$$V_y = V_c \cdot \frac{I}{R} \tag{3.13a}$$

$$V_x = V_c \cdot \frac{J}{R} \tag{3.13b}$$

Again, we can define maximum machine displacements such that

$$R_{max} = I_{max} = J_{max}$$

Using these expressions, equations (3.13) can be written

$$V_y^* = V_{max}^* \cdot \frac{V_c}{V_{max}} \cdot \frac{R_{max}}{R} \cdot \frac{I}{I_{max}} \tag{3.14a}$$

$$V_x^* = V_{max}^* \cdot \frac{V_c}{V_{max}} \cdot \frac{R_{max}}{R} \cdot \frac{J}{J_{max}} \tag{3.14b}$$

Recalling that starred terms (e.g., V_x^*) represent pulse rates, the above expression can be schematically represented as shown in Figure 3.15.

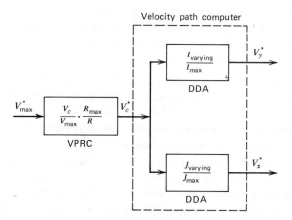

Figure 3.15 Pulse rate generation for circular path interpolation. (From Ertell, G., *Numerical Control*, copyright © 1969 by John Wiley & Sons. Reproduced with permission.)

In a manner similar to the path circuitry for linear interpolation, a pulse rate corresponding to the maximum machine feedrate is fed into a VPRC which outputs V_c^*. The path velocity pulse rate, V_c^*, is passed through two circuits which effectively multiply V_c^* by a uniformly varying quantity. The circuit which produces a uniformly time varying value is called a *digital differential analyzer* (DDA) and in its simplest form consists of *up* and *down* integrators, an adding mechanism called the *adder control*, and switching (*gating*) circuitry (Reference 3).

Each integrator can be represented as shown in Figure 3.15. Each time a velocity pulse activates the adder control, a number which has been preset into the addend register is added to the contents in the accumulator register. In this way constantly increasing or decreasing values of I and J can be generated.

The path resulting from circular interpolation is not a perfect circular arc but is an excellent incremental approximation to the arc. An NC machine with both linear and circular interpolation capabilities can follow complex contours with accuracy, but for specialized applications in which *sculptured surfaces* are machined, extended interpolation methods are available.

Parabolic Interpolation.　A parabolic interpolation system generates intermediate points along a parabolic path which has been defined either by three points in a plane (not necessarily the principle planes) or by two end slopes. The principle advantage of *parabolic*, as opposed to *circular*, interpolation is that a single parabola can sometimes be made to fit a curve that would require a number of circular arcs. Machines with a parabolic interpolation capability require fewer NC data blocks to follow rapidly changing contours.

The mathematical expression defining the parabolic path can be simulated by analog circuitry in a manner similar to methods used for linear and circular interpolation. To maintain a constant feedrate on the parabola, the more complicated equations of a general parabola must be solved in terms of the vector components of velocity.

Cubic Interpolation.　We have discussed three traditional interpolation systems which were developed for use with *hardwired* circuitry within the MCU. Cubic interpolation is available on NC machines in which part of the control function is taken over by a programmable minicomputer. Such machines are called computer numerical control systems (CNC).

In CNC systems, the path computer is a software program in the control computer which develops the coefficients for a cubic spline (see Chapter Eight) through an arbitrary number of specified points. The resultant curve has a continuous slope; that is, it is free of discontinuities which would occur if more than one arc or parabola were required to define the contour.

The mathematical theory of cubic splines is more complex than the equations for the other interpolation methods; however, the use of splining techniques, either as an interpolation function or as a means of generating the end points for line segments, is the most accurate method for contour approximation.

3.4.3 Contouring Path Velocity—Additional Considerations

Although the path computer circuitry generates commands that should produce a near perfect path, mechanical and control system characteristics may cause the resultant path to deviate from the theoretical trajectory. In general, the following velocity relationship must be maintained:

$$\sqrt{V_x^2 + V_y^2 + V_z^2} = \text{constant}$$

where the constant is the desired feedrate. For contouring systems that only approximate the above relationship, input velocities are corrected so that drift error is minimized.

Two methods of trajectory correction are commonly used. The first method of correction, *pursuit control*, uses the angle between a position, P, and the endpoint of a line segment, P_2, to correct the velocity, that is,

$$V_y / V_x = \tan \phi$$

Referring to Figure 3.16, the current point is corrected to move directly toward the endpoint, P_2. The second method, called *trajectory control*, attempts to bring the current position onto the ideal path P_1P_2 (Figure 3.17a). Trajectory control provides the more elegant form of path correction; however, changes in control system characteristics can cause the return to be too fast or too slow (Figures 3.17b, c). To maintain high path accuracy, methods such as pursuit and trajectory control are invoked at intervals of a few hundredths of a second.

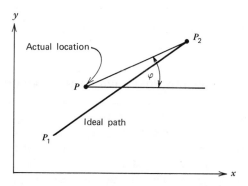

Figure 3.16 Path correction using pursuit control.

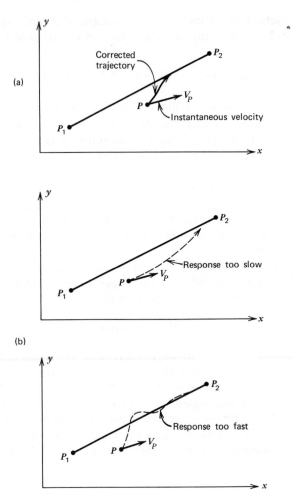

Figure 3.17 Path correction using trajectory control.

3.5 Differences Between Positioning and Contouring Systems

The most significant difference between positioning and contouring NC systems is in the design and function of their respective machine control units. Contouring systems contain path computer circuitry not found in positioning systems, and a contouring control system consumes data at a rate that can be a thousand times as rapid as its positioning counterpart.

Other important areas of difference are that whereas the positioning control can generally use relatively low power and slow response servo components, a contouring servo component must have high power, rapid response, and linearity. These characteristics are required because the system must accurately control two or more moving axes. The result of these factors is that contouring controls are intrinsically more expensive, complex, and difficult to maintain than positioning systems, but provide a much more sophisticated path following capability.

3.6 Analysis of a Typical NC System

The analysis of an NC machine tool may be undertaken in a manner that considers each of the major system components independently. Since the total system response is a function of the individual component transfer functions, the power amplifier, servomotor, and machine tool slide itself can be termed the *actuation components*. These components are controlled by an analog signal which is supplied by a *filtering system* comprised of an error detector, digital to analog converter, and control amplifiers. Finally, the error signal is provided within the feedback loop by means of a transducer attached to the servomotor output or to the tool slide. Such a system is illustrated in Figure 3.5.

3.6.1 Command Signals and Error Detection

The command signal for an NC machine is an electronic pulse corresponding to the resolution of the machine's positioning accuracy. For example, if a pulse is generated every 1/30,000 second, and the positioning accuracy is 0.002 mm, then the maximum feedrate is 60 mm/sec. The command signal enters a phase difference detector which compares the input signal to the transducer feedback signal. This produces an error signal which is passed through the filter network.

The output position of the machine table is directly related to the number of revolutions made by the leadscrew. A rotary transducer* mounted on this screw produces an output signal of the form $A \sin(\omega t + \theta_0)$, where ω is a constant frequency and θ_0 is a phase shift which is a measure of the leadscrew phase angle. This sinusoidal output is transformed to a square wave as illustrated in Figure 3.18. If the rotary transducer frequency is $\omega = 200$ Hz, then from the figure, $t_0 = 1/200$ second.

The input command is also transformed into a square wave with a frequency $\omega = 1/t_0$ and a phase shift of $\theta_c t_0/2\pi$ where θ_c is a function of the input position command (Figure 3.19). The phase difference detector takes each signal (i.e.,

---

* The rotary transducer described here is called a resolver.

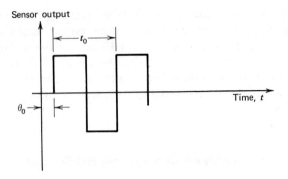

Figure 3.18 Transducer output.

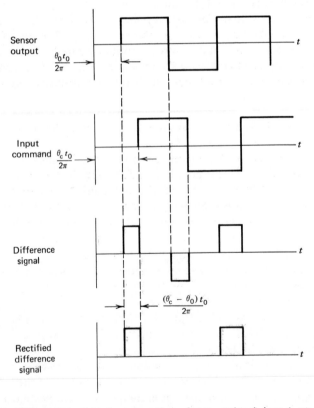

Figure 3.19 Difference signal generation from pulsed input and transducer signals.

the command signal and the transducer signal) and produces a square wave of the difference. This square wave is rectified to produce a *difference signal*, $x(t)$.

The rectified difference signal, $x(t)$, is applied to a filter network. Consider a first order filter with a transfer function of the form

$$G_f(s) = \frac{1}{1 + \tau s} \tag{3.15}$$

where τ is a time constant.

Using Laplace transform methods, it can be shown that for a single pulse input, where $\theta_c = 0$ and θ_0 is constant, the output from the filter is

$$y_1(t) = e^{-t/\tau} [e^{\theta_0 t_0/2\pi\tau} - 1]$$

for values of $t \geq \theta_0 t_0 / 2\pi$.

Since each pulse will produce output that is identical to $y_1(t)$, except for the time value,

$$y_n(t) = \sum_{k=0}^{n-1} e^{-(t + kt_0)/\tau} [e^{\theta_0 t_0/2} - 1] \tag{3.16}$$

where $y_n(t)$ is the response to n successive pulses.

A general expression for the output as a function of n pulses can be written as

$$y_n(nt_0) = \frac{1 - e^{-\theta_0 t_0/2\pi\tau}}{1 - e^{t_0/\tau}} [1 - e^{-nt_0/\tau}] \tag{3.17}$$

For most systems θ_0 and t_0 are much smaller than τ; hence, equation (3.17) can be rewritten as

$$y_n(nt_0) = \frac{\theta_0}{2\pi} [1 - e^{-nt_0/\tau}] \tag{3.18}$$

where $\theta_0/2\pi$ is the gain of the system.

Because the increments, t_0, are small, we can write equation (3.18) as a continuous function

$$y(t) = \frac{\theta_0}{2\pi} [1 - e^{-t/\tau}] \tag{3.19}$$

Equation (3.19) illustrates the continuous time response which is elicited through the filter described in equation (3.15). The pulsed input to any system element with a known transfer function can be evaluated in a similar manner.

3.6.2 Pattern Errors in a Contouring System

Pattern errors occur when the servo components that control different axes of an NC machine are slightly mismatched, that is, when they have slightly different transfer functions or gain characteristics. At any given time, pattern error is defined as the minimum distance between the contour that is cut and the contour that is desired.

Suppose that a straight line is to be cut in a plane. Velocity components for feedrate are v_1 and v_2 in their respective axes, and the axis steady state servo errors are e_1 and e_2, respectively. The pattern error, E, is specified as shown in Figure 3.20.

To develop an expression for the pattern error, E, for the two-dimensional case illustrated, consider the *effective time lag*, λ, which represents the difference between the actual position and the command position. The vector components for the path can be represented as shown in Figure 3.21. From the figure

$$\mathbf{E} = \mathbf{F} + \mathbf{e}_1 + \mathbf{e}_2 \qquad (3.20)$$

Recalling that the path direction has the velocity components v_1 and v_2,

$$\mathbf{F} = (\lambda v_1, \lambda v_2)$$

where $(\lambda v_1, \lambda v_2)$ represent the scalar components for the first and second axes.

For a minimum pattern error, $E = 0$. Therefore, from equation (3.20)

$$0 = \lambda^2(v_1^2 + v_2^2) - \lambda v_1 e_1 - \lambda v_2 e_2$$

and, solving for λ, the effective time lag can be written

$$\lambda = \frac{e_1 v_1 + e_2 v_2}{v_1^2 + v_2^2} \qquad (3.21)$$

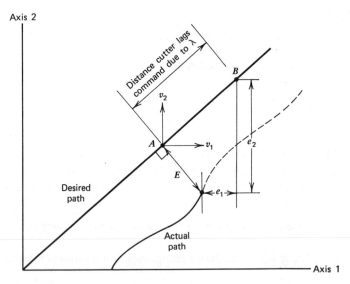

Figure 3.20   Pattern error in a two-axis system.

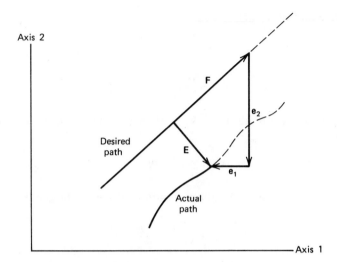

Figure 3.21   Vector definition for pattern error analysis.

From Figure 3.21 it can be seen that

$$\mathbf{E} = (\lambda v_1 - e_1, \lambda v_2 - e_2)$$

Substituting equation (3.21) into the above expression, we get

$$\mathbf{E} = \left(\frac{e_1 v_1^2 + e_2 v_1 v_2}{v_1^2 + v_2^2} - e_1, \frac{e_1 v_1 v_2 + e_2 v_2^2}{v_1^2 + v_2^2} - e_2\right)$$

or

$$E = \frac{|v_2 e_1 - v_1 e_2|}{\sqrt{v_1^2 + v_2^2}} \tag{3.22}$$

the magnitude of the pattern error in two dimensions.

Furthermore, it can be shown (Reference 4) that for an n-dimensional system the pattern error can be expressed as

$$E = \left\{ \sum_{j=1}^{n} \left[\frac{v_j \sum_{i=1}^{n} e_i v_i - e_j \sum_{i=1}^{n} v_i}{\sum_{i=1}^{n} v_i^2} \right]^2 \right\}^{1/2} \tag{3.23}$$

Servo Mismatch.   In a linear system, the servo errors are proportional to the respective axis feedrates:

$$e_1 = g_1 v_1; \; e_2 = g_2 v_2$$

where g_1 and g_2 are the transfer functions (gains) of the servo components.

Then, from equation (3.22),

$$E = \frac{|v_2 g_1 v_1 - v_1 g_2 v_2|}{\sqrt{v_1^2 + v_2^2}}$$

$$E = \frac{|v_1 v_2|}{\sqrt{v_1^2 + v_2^2}} |g_1 - g_2| \qquad (3.24)$$

If the maximum axis feedrate is V_{max}, then the maximum pattern error will occur when $v_1 = v_2 = V_{max}$:

$$E_{max} = \frac{V_{max}}{\sqrt{2}} |g_1 - g_2|$$

Therefore, should we wish to allow a pattern error no larger than E_{max},

$$|g_1 - g_2| \leq \frac{\sqrt{2} E_{max}}{V_{max}} \qquad (3.25)$$

If the average gain for components in each axis is g then the maximum deviation from normal must be Δg, and from the above expression let

$$|g_1 - g_2| = (g_1 - g) - (g_2 - g) = 2\Delta g$$

Therefore, equation (3.25) can be rewritten

$$\frac{\Delta g}{g} = \frac{E_{max}}{g V_{max} \sqrt{2}} \qquad (3.26)$$

and expresses the percent difference in gain (mismatch) as a function of pattern error.

Suppose that the maximum allowable pattern error is 0.002 mm for a maximum feedrate of 1 mpm. Nominal gain value is 1/60 second and from equation (3.26)

$$\frac{\Delta g}{g} = 0.005$$

That is, the two axes must have characteristics that match to within an accuracy of 0.50 percent.

3.6.3 System Performance and the Time Constant, τ

Let us examine the dependence of the system performance on the value of the filter network time constant, τ. As indicated by the condensed block diagram illustrated in Figure 3.22, the NC machine tool block (i.e., the power amplifier, actuation components, and velocity feedback) can be represented by the transfer function

$$G_{nc}(s) = \frac{1}{(1 + \tau_1 s)\left(1 + \frac{2\beta}{\omega_n}s + \frac{1}{\omega_n^2}s^2\right)} \tag{3.27}$$

Equation (3.27) is derived using the same type of analysis as found in Section 2.6.* In the equation, ω_n is the natural frequency of the system, τ_1 is a component time constant, and β is a damping term.

The filter network transfer function is obtained from equation (3.15), and the transfer function for the position feedback loop is

$$G_f(s) = 1 + \tau_2 s$$

where τ_2 is the transducer time constant.

Because the second order terms in equation (3.27) make a negligible contribution to the system performance, the quadratic term may be eliminated. The simplification yields a block diagram of the form shown in Figure 3.23.

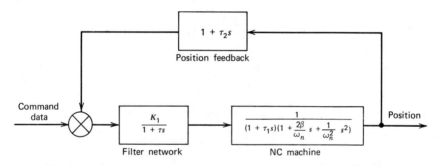

Figure 3.22 NC system block diagram with typical transfer functions.

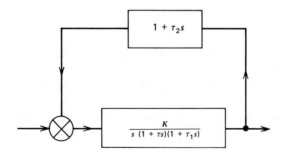

Figure 3.23 NC system block diagram simplified by removal of second order characteristics.

* The reader is urged to review this material, if necessary.

Using the relationships developed in Chapter Two, the open loop transfer function for the system illustrated in Figure 3.23 is

$$G_L(s) = \frac{K(1 + \tau_2 s)}{s(1 + \tau s)(1 + \tau_1 s)}$$

where K is a combination of constants.

Recalling that s is a complex number, for sinusoidal input we can write

$$G_L(s) = \frac{K(1 + j\omega\tau_2)}{j\omega(1 + j\omega\tau)(1 + j\omega\tau_1)} \tag{3.28}$$

It is evident from equation (3.28) that for any frequency, ω, the value of the transfer function, and, hence, of the loop gain, decreases as the time constant, τ, increases.

The *critical time constant*, τ_c, is defined as the value of τ for which $G_L(j\omega)$ *is* 1. Since the magnitude of $G_L(s)$ can be expressed as its absolute value

$$\frac{K|1 + j\omega\tau_2|}{\omega|1 + j\omega\tau_c| \cdot |1 + j\omega\tau_1|} = 1$$

or, solving for τ_c,

$$\tau_c = \frac{1}{\omega}\left[\frac{(1 + \tau_2^2\omega^2)K^2}{(1 + \tau_1^2\omega^2)\omega^2} - 1\right]^{1/2} \tag{3.29}$$

Since the input signal is sinusoidal in nature, the output will take the form

$$y(t) = G_L(\omega, \tau)C \sin(\omega t + \theta)$$

where G_L is the transfer function and C is a constant. Hence, the overall system output depends both on frequency and filter time constant. From the previous discussion of system response (Chapter Two), it is clear that as τ increases the response of the system slows down.

3.6.4 System Bandwidth and Cornering Error

Bandwidth has been previously defined as the frequency range over which there is an appreciable system response to a given input. It can be shown that as the time constant, τ, increases, the system bandwidth decreases. One effect of a smaller bandwidth is the inability of an NC machine to trace a perfect 90° corner. Figure 3.24 illustrates this problem.

The closed loop transfer function for each of the axes of an NC machine tool can be approximated by

$$G_{CL}(s) \approx \frac{1}{1 + s/B}$$

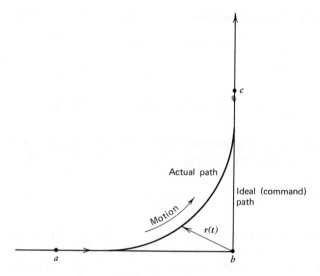

Figure 3.24 Cornering error.

The inverse Laplace transform* of the above expression can be written:

$$\mathscr{L}^{-1}[G_{CL}(s)] = \mathscr{L}^{-1}\left[\frac{1}{1/B(s-(-B))}\right] = Be^{-Bt} \qquad (3.30)$$

Now, an expression for the cornering error, r, in terms of the bandwidth, B, can be established. Considering the ideal path, abc, in Figure 3.24, a cutter moving in the direction shown, at a feedrate F, would be at the command position indicated with respect to time:

$$
\begin{aligned}
t < 0; &\quad x_c(t) = Ft, &\quad y_c(t) = 0 \\
t \geqslant 0; &\quad x_c(t) = 0, &\quad y_c(t) = Ft
\end{aligned}
\qquad (3.31)
$$

where t is measured positive at the point of direction change.

The actual path coordinates, that is, the response, must take on an exponential form based on the nature of the transfer function (equation 3.30). Hence, for $t \geqslant 0$ it can be shown that

$$x_a(t) = \frac{F}{B}e^{-Bt}$$

$$y_a(t) = F\left[t - \frac{1}{B}(1 - e^{-Bt})\right] \qquad (3.32)$$

* See Chapter Two for details.

To determine the minimum error

$$r(t) = \{[x_a^2(t) + y_a^2(t)]^{1/2}\}_{min}$$

Using equations (3.32) a minimum expression becomes

$$r(t)_{min} = \frac{R}{B}\{[e^{-2Bt} + (Bt - 1 + e^{-Bt})^2]^{1/2}\}_{min}$$

The minimum value of the quantity in brackets for any given B is found to be 1/2. Hence, the minimum cornering error for an NC machine with bandwidth B is

$$r \approx \frac{F}{2B} \tag{3.33}$$

A more general expression for cornering error can be written when individual servo characteristics are considered (Reference 5). The developed model is based on two assumptions: (1) the force supplied by the servomotor must accelerate machine members and overcome viscous friction; and (2) cutting tool lag is directly proportional to the instantaneous force.

Expressing the first assumption mathematically,

$$M\frac{d^2x}{dt^2} = F_x - k_f\frac{dx}{dt}$$
$$M\frac{d^2y}{dt^2} = F_y - k_f\frac{dy}{dt} \tag{3.34}$$

where M is the effective accelerated machine tool mass, and k_f is a functional constant that includes servo damping.

The second assumption can be written

$$F_x = k_s(x_c - x)$$
$$F_y = k_s(y_c - y) \tag{3.35}$$

where k_s is the servo constant and x_c, y_c are the command position coordinate values.

Since this analysis considers any corner angle, θ, the command position with respect to time can be written as

$$\left.\begin{array}{l} x_c(t) = Ft\cos\theta \\ y_c(t) = Ft\sin\theta \end{array}\right\} t \geq 0 \tag{3.36}$$

Using equations (3.34) through (3.36) the following general expressions for actual cutter position can be derived:

$$x_a(t) = e^{c_1 t}[k_1 \cos C_2 t + k_2 \sin C_2 t] + F\left(t - \frac{k_f}{k_s}\right)\cos \theta$$

$$y_a(t) = e^{c_1 t}[k_3 \cos C_2 t + k_4 \sin C_2 t] + F\left(t - \frac{k_f}{k_s}\right)\sin \theta$$

(3.37)

where C_1, C_2, and k_i, $i = 1$ to 4, are constants.

Equations (3.37) have been shown to agree quite well with actual machine movements. In these equations, the time constant of the system is implicitly expressed in the machine servo constant, k_s.

References

1. Anand, D. K., *Introduction to Control Systems*, Pergamon Press, New York, 1974, pp. 240–52.
2. Bezier, P., *Numerical Control—Mathematics and Applications*, Wiley, New York, 1970, pp. 13–19.
3. Ertell, G. G., *Numerical Control*, Wiley-Interscience, New York, 1969, pp. 100–22.
4. Bekey, G. A., Nahi, N. E., and Payne, H. J., "Evaluation of Control Systems for Numerically Controlled Machine Tools," *IBM report*, 1968 (approx.).
5. Burkley, R. M., and Broadwell, W. B., "Dynamic Model for Contouring NC Devices," *Proceedings of the Ninth Annual Meeting and Technical Conference*, Numerical Control Society, Chicago, 1972, pp. 156–68.

Problems

1. Develop a schematic diagram for a three-axis NC machine showing input and buffer storage, all servomotors, and the feedback loops for each axis. How many storage registers are required if the machine uses a positioning controller?

2. An NC machine is capable of executing 10 blocks of NC data per second. Each block contains an average of 30 characters of information, and internal data transfer between registers occurs at a rate of 1 character every 100 μsec. Given a system which uses buffer and active storage registers, could a reader with a speed of 100 characters/sec be used with the above configuration?

3. The control unit of an NC machine generates 42,380 pulses as it moves along a principle axis. If 1 pulse = 0.002 mm and the pulses are generated in 8.5 sec, what is the feedrate?

4. Discuss open and closed loop feedback systems for NC machines. Abstract two outside sources (e.g., engineering journals, periodicals) that compare these methods.

5. (a) For the control system block diagram in Figure P.3.1 calculate the sensitivity of the system to a change in the feedback transducer characteristic, K_4. (b) How will changes in the actuation devices block, K_3, affect the sensitivity of the system?

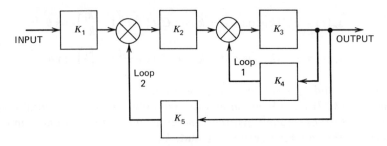

Figure P.3.1   Problems 3.5, 3.6, and 3.7.

6. Over a given range of operating conditions the characteristic of the velocity transducer varies by ± 0.5 percent. Over the same operating range the characteristic of the actuation devices (see Figure P.3.1), K_3, varies by ± 1.5 percent.
    (a) Taken individually, how will each variance affect the entire NC machine?
    (b) Taken together, how will the variance in Loop 2 affect the sensitivity of the NC machine?
    (c) The total transfer function for the NC machine must vary by no more than ± 0.6 percent. Is the characteristic variance discussed above tolerable?

7. In Figure P.3.1, what component block would seem to have the greatest effect on system sensitivity? Justify your answer.

8. A manually operated milling machine has been found to have severe backlash in the transverse direction. Is there a method for positioning the table (using the standard controls) that would keep the effect of backlash to a minimum? Can you draw any general conclusions about positioning and backlash?

9. Describe the amplifier dead-zone concept in your own words. From Figure 3.9 in the chapter, determine the smallest magnitude of the dead-zone for a machine in which 1 pulse $= 0.002$ mm.

10. An NC positioning system uses an MCU that evaluates the contents of the coordinate registers simultaneously. Given $(0, 0)$ as a starting point, draw the tool path if the registers contain the following contents in sequence:

x-register	y-register
1.0	1.0
1.5	3.0
4.5	3.0
6.5	4.0
1.0	1.5
0.0	0.0

11. (a) Using the table of coordinates from Problem 10, draw the tool path for a machine with an MCU that evaluates one register at a time.
    (b) Draw the tool path for an MCU with linear interpolation.

12. What type of velocity control would be required for a positioning system that can tolerate no overshoot? What type of function can be used to approximate the control response?

13. The motion illustrated in Figure P.3.2 is to be executed by an NC machine tool. If the maximum pulse rate is 30,000 pulses/sec and the maximum linear motion is 0.6 m, what is the pulse rate output for each axis? The desired feedrate is 1.2 mpm.

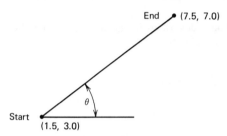

Figure P.3.2   Problem 3.13.

14. A 45° circular arc is to be machined at a feedrate of 2 mpm. Assuming that the arc radius is the maximum allowable, what are the pulse rate outputs when $\alpha' = 15°$? See Figure P.3.3 for details.

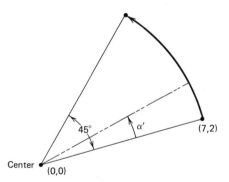

Figure P.3.3   Problem 3.14.

15. Design a *spiral arc* interpolation system for a hypothetical machine that takes cylindrical coordinate input.
    (a) Discuss velocity considerations.
    (b) Indicate input to the system.
    (c) Develop and describe a velocity path computer that is analogous to those described in Section 3.4.
    Recall that in cylindrical coordinates (r, θ, z) $x = r \cos \theta$, $y = r \sin \theta$, and $z = z$.

16. Using the expression developed for pattern errors in an n-dimensional space, equation (3.23), express the pattern error, E, for a three-axis NC machine given the feedrate components v_1, v_2, v_3 and steady state servo errors e_1, e_2, e_3.

17. From the expression developed in Problem 16, find an expression that specifies the percent mismatch for a three-axis system.

18. If the servo components of a two-axis NC machine are known to have a mismatch of 0.45 percent, and a nominal gain of 1/60 sec, what will be the maximum pattern error at the maximum feedrate of 3 mpm?

19. From the expression derived in Problem 17, determine the percent accuracy required for a 3-D system with a nominal servo gain of 1/60 sec and an allowable $E_{max} = 0.02$ mm, and $V_{max} = 2$ mpm.

20. Derive equations (3.37) based on the model described in the text. Describe the significance of each of the constants in the equation.

Chapter Four
NC System Components

The numerical control system is comprised of a number of hardware subsystems interconnected with a communications interface and controlled by the machine control unit. These NC subsystems can be grouped into the following categories:

1. MCU digital circuitry.
2. Feedback hardware.
3. Actuation systems.
4. Power amplification systems.

Figure 4.1 illustrates the basic division of hardware in an NC system. Electronic circuitry is essential for the logical, memory, and computational operations performed by the control system. Signal amplification is required to transform the low voltage output of the MCU to a signal powerful enough to drive the system. Drive components of suitable design accurately position and power the machining operation.

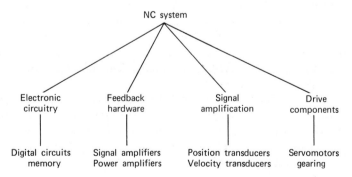

Figure 4.1 Elements of an NC system.

It is difficult to assign degrees of importance to individual NC components since performance depends upon the proper functioning of all parts. However, closed loop NC systems rely upon information feedback to initiate error correction and provide accuracy in tool velocity and position. Transducers initiate this process.

4.1 Transducers for NC Machines

4.1.1 Transducer Types

All closed loop NC systems must receive accurate information concerning the instantaneous position of the controlled component. Other information, such as velocity, direction of movement, and, sometimes, acceleration, is also desirable. Although classical devices* suitable for the measurement of these quantities are available, special transducers have been devised for use with NC machine tools.

Three basic types of transducers are generally used:

1. *Numerical encoding devices.*
2. *Pulse summation devices.*
3. *Analog devices.*

It is important to note that many NC systems use a combination of transducer types. The choice depends on the quantity to be measured, the location of the measuring device, feedback data type, and the desired accuracy.

4.1.2 Linear Transducers

A linear transducer is attached to the NC machine in a manner that enables it to *read* a special scale attached to a moving table. Pulse generating linear transducers were discussed in Chapter Three. Variable reluctance linear transducers are also used. In these units the stationary transducer contains an induction coil to which an ac voltage is applied. The moving scale is separated from the transducer by a small gap.

The grid pattern of the transducer matches the pattern on the movable scale, and therefore, a full cycle of voltage occurs as the relative position between the patterns changes by one grid. Each voltage cycle is effectively a pulse which can be counted to indicate relative displacement. Such linear devices can be supplied with accuracies to 0.004 mm.

* For example, LVDTs, tachometers, and accelerometers.

4.1.3 Rotary Transducers

The rotary transducer is the most common sensing device for NC machines. Its popularity can be attributed to compact size and the flexibility with which it can be used. Figure 4.2 illustrates common rotary transducer measurement configurations.

There are two primary types of rotary transducers used in modern NC equipment: *resolvers* and *encoders.* The resolver is an analog device whose output is converted to digital form, whereas the encoder is a numerical device which outputs digital data directly.

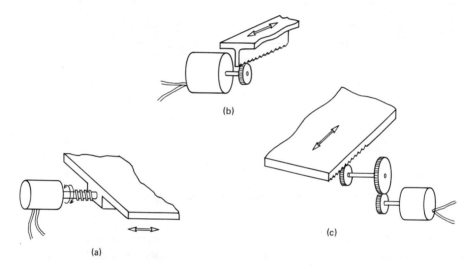

Figure 4.2 Rotary transducer configurations: (a) Direct connection to leadscrew. (b) Rack and pinion connection. (c) Geared connection.

4.1.4 The Resolver

The resolver, also called a *synchro*, consists of an assembly which resembles a small electrical motor with a stator-rotor configuration. The stator contains two coils excited by voltages of equal amplitude but 90° out of phase. As the rotor turns, the phase relationship between the stator and rotor voltages corresponds to the shaft angle, so that one electrical degree of phase shift corresponds to one mechanical degree of rotation.

Thus, the phase shift between the rotor and stator coils of a resolver is an exact measure of the input shaft rotation. Resolvers with accuracies of two to three seconds of arc are not uncommon.

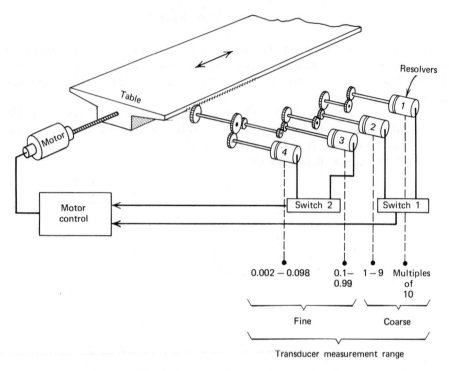

Figure 4.3   A four resolver positioning system.

In many NC applications, resolvers are used in tandem to provide a method for the elimination of position ambiguity which results from multiple revolutions. Such systems provide improved accuracy over a broad range of motion.

Referring to Figure 4.3, it is seen that position control can be accomplished using *coarse* and *fine* feedback originating from different rotary transducers in the multiply geared arrangement. That is, the input command to the control unit is divided into a coarse command (i.e., motion to the nearest millimeter) and a fine command (motion to the nearest 0.002 mm). Until the coarse command has been satisfied, Switch 1 is open and feedback is returned from transducers 1 and 2. Once the coarse command has been satisfied, Switch 1 is closed and Switch 2 is opened so that feedback originates from transducers 3 and 4.

4.1.5 Encoders

The rotary encoder produces digital output which corresponds to a special pattern contained on a rotating disk within the device. Two basic types of rotary encoders are found in NC machines: the *numerical encoder* and the *pulse generating encoder.*

A numerical encoder generates the numerical value of the measured position by interpreting and transmitting a code based on the printed disk pattern. To illustrate the use of the numerical encoder consider the binary coded* disk, shown in Figure 4.4*a.*

The darkened areas on the disk, when evaluated along any radius of the circle, represent unique binary numbers. The coded areas can be *read* using photoelectric, magnetic, or electrical conducting systems. It can be seen by

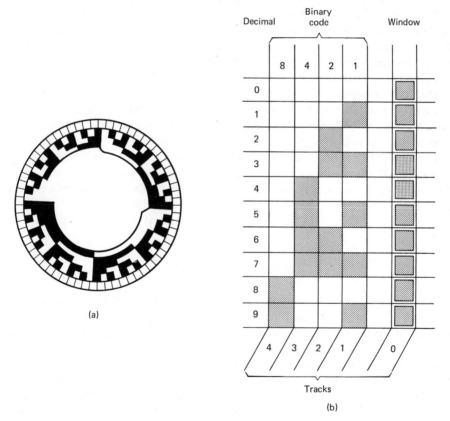

(a)

(b)

Figure 4.4 A numerical encoder disk. (a) Binary code pattern on disk. (*b*) Magnified view of disk section. (From Olesten, N.O., *Numerical Control*, copyright © 1970 by John Wiley & Sons. Reproduced with permission.)

* For a discussion of binary codes, the reader is referred to Chapter Six.

referring to a segment of the disk magnified in Figure 4.4b that each track sets a binary digit *on* or *off* depending on the presence or absence of the pattern at the specific track position.

Regardless of the reading technique, there will always be some small angle between values for which adjacent patterns may be read concurrently. This will produce an ambiguous and erroneous code. Consider the pattern illustrated in Figure 4.4b. Assume that the encoder reading heads have just changed from decimal 3 to decimal 4. If at the transition point the binary areas 2 and 1 were extended by 0.01 mm, the encoder output would be 0111 (decimal 7) rather than 0100 (decimal 4). Although this erroneous reading would last for only a fraction of a second, incorrect data could be generated.

To remedy the above problem *antiambiguity logic* is incorporated into the encoder. One technique is called *window code*. As each radial pattern passes the reading heads, reading is permitted only while the window (see Figure 4.4b) in the outermost track is *on*. Hence, by suppressing evaluation at the pattern transition line erroneous data can be eliminated.

The *pulse generating encoder* is effectively the rotary analog of the linear pulse generating transducer. Like the numerical counterpart, the pulse generating encoder can use an electric, photoelectric, or magnetic reading technique. Consider a photoelectric configuration as an example.

The encoder produces pulsed output which is generated as a disk containing finely etched lines rotates between an exciter lamp and one or more photo diodes. A square wave is generated each time a line breaks the light circuit. The total number of pulses generated in a single revolution is a function of the number of lines etched on the disk. Typical disks may contain from 50 to 2500 lines.

4.1.6 Transducer Placement

A transducer must be located in a manner that provides the greatest possible accuracy. The most common methods for transducer placement are: (1) direct connection to the leadscrew (rotary), (2) precision rack and pinion arrangement (rotary), and (3) linear measuring devices.

Transducers connected directly to the leadscrew of moving components (refer to Figure 4.2a) require no additional gearing, and, from a mechanical standpoint, constitute the simplest arrangement. However, the transducer must be adjusted to compensate for any error in screw threads, and it must be placed on the end of the leadscrew subjected to the smallest torsional load.

To improve the precision of feedback, the transducer can be connected in a manner that makes it independent of the drive system. The precision rack and pinion arrangement (Figure 4.2b) measures the movement of the table in a more direct manner. Provided the rack and pinion gearing is precise and gear backlash has been eliminated, the measurement technique produces extremely accurate feedback.

4.1.7 Transducer Signal Processing

The output of a transducer must be processed in a manner that will enable the MCU to compare the feedback signal to the command signal and take appropriate action. As an example, consider the resolver.

The resolver works with sinusoidal voltages. Therefore, the voltage supply to the resolver is synchronized to the digital system, and the resolver output is converted from sinusoidal to digital form.

The MCU clock supplies square wave (digital) pulses which are processed by a *shaper circuit* into two equal amplitude sinusoidal signals; each out of phase by 90°. These signals are used to excite the stator winding of the resolver. The resolver rotor generates a sinusoidal output signal which is reconverted into a square wave by an electronic *clipping circuit*.

The next step is to develop a method for the comparison of command and feedback signals. Referring to Figure 4.5, the position error is the phase difference between the command and the feedback signals. The *comparator circuit* detects whether the phase *leads* or *lags* that of the command signal. If a difference is found, the comparator attempts to compensate for the difference by generating an error signal of appropriate polarity.

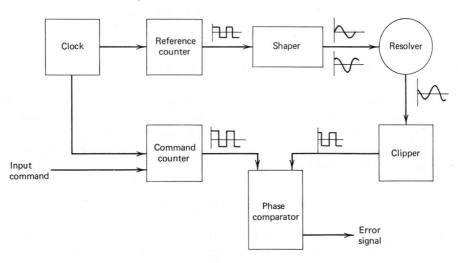

Figure 4.5 Transducer signal processing.

4.2 NC Actuation Systems—an Overview

The commands which control the motion of an NC machine are transformed into mechanical work by the NC *actuation system*. The actuation system consists of three major elements: the power amplifier, drive motor(s), and drive transmission devices.

Input to the actuation system is in the form of low level signals that have resulted from a comparison of command and feedback in a closed loop system. Usually the power of these low level signals is not sufficient to drive the NC servomotors, but a power amplifier is used to overcome this deficiency.

The power amplifier is often required to effect both energy and power conversion as the input and output from the device may be in different forms and levels. An example of this is a hydraulic power amplifier that requires an input of a low level electrical signal, but produces an output of high pressure hydraulic flow capable of driving a servomotor. The power amplifier is therefore the link between the control system and the servo drive components.

Servomotors for numerical control machines transform the power amplified command signal into work which produces motion. A servomotor should have low rotor inertia for rapid speed changes, a wide range of operating speeds in both directions of rotation, and stable speed versus torque characteristics. Its response should be linear over a wide range of operating conditions.

The power transmission components of the actuation system depend on the type of system in use (e.g., electrical or hydraulic) and on the motion requirements. All components are precisely machined to avoid the mechanical problem of backlash. Three types of actuation systems are commonly found in . NC equipment: (1) electromechanical, (2) hydraulic, and (3) pneumatic.

4.3 Electromechanical Actuation Systems

Electromechanical actuation systems make use of many combinations of motors, power amplification devices, and drive components. Two common configurations—open loop positioning systems using the stepping motor and closed loop positioning or contouring systems requiring power amplification for a dc motor—are considered.

4.3.1 The Stepping Motor

The stepping motor is a device that converts electrical pulses into a proportionate mechanical response. Because it accepts digital input, this motor is ideally suited to open loop systems in which precise motion must be executed for each command pulse. The stepping motor is generally limited to applications requiring low torque output. For NC, an *electrohydraulic* stepper configuration, consisting of a stepping motor controlling a hydraulic power amplifier, can be used to generate high output torque.

Figure 4.6a illustrates a simplified cross section of a *variable reluctance* stepping motor. A ferromagnetic multipole rotor moves inside a magnetic field generated by the wound stator. When an individual stator pole is energized, the rotor moves to a stable equilibrium position corresponding to the minimum reluctance of the magnetic circuit; that is, a rotor pole aligns itself with the energized stator face to face. As the stator poles are energized in sequence, the rotor rotates by angular steps as the equilibrium is maintained. The number of angular positions can be doubled if two stator poles are energized simultaneously. In this case the equilibrium position falls between the two stator poles.

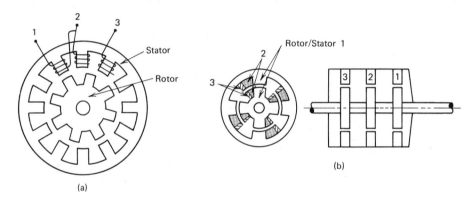

Figure 4.6 Stepping motor configurations: (a) Single rotor type. (b) Multiple disk configuration.

A different variable reluctance stepping motor configuration is illustrated in Figure 4.6b. The rotor consists of three to five disks having the same number of poles as the stator. Given n disks, each with poles separated by pitch, p, and mounted so that the rotor poles are rotated an amount p/n relative to the preceding disk, when the stators are excited in sequence, the rotor will advance by p/n for each pulse.

4.3.2 The dc Motor

In the early days of numerical control, the direct current (dc) motor played a minor role in machine drive applications because of inadequate motor control and unacceptable size to horsepower ratios. In contrast, the modern dc motor provides high torque, high response output from units one fifth the size of earlier models.

Consider two important characteristics of dc servomotors: (1) motor speed varies linearly with input voltage, and (2) motor torque varies linearly with input current. To maintain constant speed under varying loads, the input voltage to the dc servomotor must be closely controlled. The *continuous torque rating* of an electric motor indicates the continuous operating load for which motor temperature can be held at acceptable levels. The dc motor can, however, operate at torque levels well above the continuous torque rating for brief periods of time.

A typical dc motor configuration is illustrated in Figure 4.7. Readers who are unfamiliar with the basic operating principles of electric motors are referred to References 1 and 2. Although the field and armature windings of a dc motor may be connected in different ways,* the motors used in NC systems are

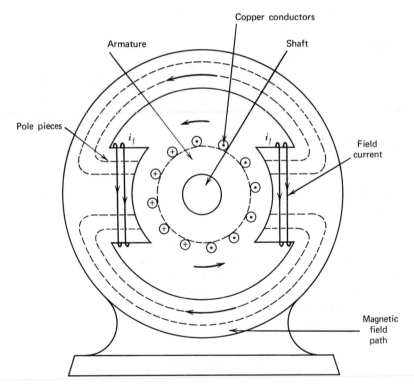

Figure 4.7   Schematic diagram of a dc motor.

* For example, shunt connected, series connected, compounded.

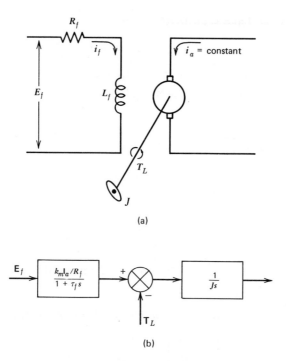

(a)

(b)

Figure 4.8 A field controlled dc motor. (*a*) Circuit schematic. (*b*) Block diagram.

generally *separately excited*, that is, using field control with fixed armature current or armature control with a fixed field.

A circuit diagram for a field controlled dc motor is given in Figure 4.8*a*. The following expression can be written for the torque, *T*, developed by the motor:

$$T = k_1 i_a \phi \qquad (4.1)$$

where i_a is the armature current, ϕ is the magnetic flux of the field, and k_1 is a motor constant.

One important characteristic for motors used in control systems is that they operate over a linear range. Referring to Figure 4.9, the flux, ϕ, increases linearly with field current, i_f, until the stator saturates. In the linear range

$$\phi = k_2 i_f$$

Then equation (4.1) becomes

$$T = k_1 k_2 i_a i_f = k_m i_a i_f \qquad (4.2)$$

If the moment of inertia of the armature is *J*, and the viscous damping coefficient is β, we have seen that*

$$T = (\beta s + Js^2)\theta + T_L \qquad (4.3)$$

where s is the Laplace operator, T_L is the load torque, and θ is the angular position of the armature (i.e., motor shaft).

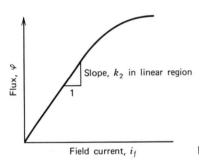

Figure 4.9 part:
Slope, k_2 in linear region
1
Flux, φ
Field current, i_f

Figure 4.9   Magnetic flux versus field current.

Referring again to Figure 4.8a, the circuit diagram for the motor yields

$$\mathbf{I}_f = \frac{\mathbf{E}_f}{R_f + L_f s} = \frac{\mathbf{E}_f}{R_f(1 + \tau_f s)} \qquad (4.4)$$

where $\tau_f = R_f/L_f$ is the *field circuit time constant*.

To determine an expression for the speed (angular velocity) of the motor, $\dot{\theta}$, we can substitute equations (4.3) and (4.4) into equation (4.2); solve for the angular position, θ, and since $\dot{\theta} = s\theta$, we obtain

$$\dot{\theta} = \frac{1}{\beta + Js}\left[\frac{(k_m \mathbf{I}_a/R_f)\mathbf{E}_f}{1 + \tau_f s} - T_L\right] \qquad (4.5)$$

Hence, given an input voltage E_f, we have developed an expression for the motor transfer function. If $\beta \ll Js$, the block diagram for the field controlled motor can be drawn as illustrated in Figure 4.8b.

The above development illustrates the manner in which the characteristic of any electric servomotor can be evaluated. It should be noted that in real applications, a constant armature current, i_a, is difficult to maintain. A constant current source can be approximated by supplying the armature with a constant voltage, V_a, in series with a large resistance (Reference 3).

---

* For further discussion, see Chapter Two.

4.3.3 Electrical Power Amplification

A detailed discussion of electronic amplifiers is beyond the scope of this book. We therefore discuss only the power amplifier as it relates to the dc motor in an NC system. The input to a power amplifier from the control system is a low level signal of high impedance which, when processed by the power amplifier, transforms the signal to low impedance suitable for energizing the servomotor.

A typical electrical power amplification system is illustrated in Figure 4.10. The external dc power source is a generator driven by an ac motor. Alternatively, the motor generator could be replaced by a controlled rectifier which converts ac line current to dc current.

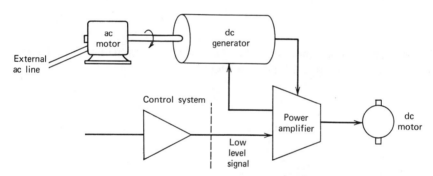

Figure 4.10 Typical electrical power amplification system.

The *silicon-controlled rectifier* (SCR) is one such solid state device. The SCR (Figure 4.11a) effectively converts an ac input signal into a series of positive pulses by *inverting* or *blocking* the negative half cycle. This process is illustrated in Figure 4.11b. A 60-Hz ac power supply will produce 60 power pulses/sec—approximating a direct current. A *full-wave* SCR doubles the number of positive pulses by reversing the polarity of the negative half of the ac cycle, as shown in Figure 4.11c.

In order to control a dc motor for NC applications, the SCR circuit must be capable of reversing rotation. A simplified half-wave reversible SCR is illustrated in Figure 4.12. Field current flows from P_1 to P_2. If SCR_1 is activated (i.e., gate 1 receives a *turn-on* signal), current flows from P_3 to P_4, causing the motor armature to turn *clockwise*. If SCR_2 is activated, current flows from P_4 to P_3 and the armature rotates in a *counterclockwise* direction.

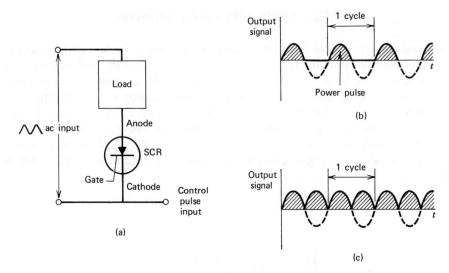

Figure 4.11 Silicon Controlled Rectifier (SCR). (a) Schematic. (b) Results of half-wave rectification. (c) Full-wave rectification.

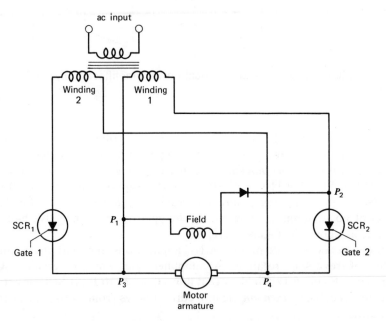

Figure 4.12 SCR motor control circuit.

4.4 Hydraulic Actuation Systems

A hydraulic actuation system consists of a *pump* to develop hydraulic pressure, a *control valve* to monitor the flow of oil, and an *actuator* such as a hydraulic motor or cylinder. Hydraulic systems are used extensively in NC equipment that requires: (1) rapid system response, (2) good performance in areas of shock or vibration and (3) high torque output.

Hydraulic actuation is used in the majority of contouring NC systems because the high force output of a hydraulic system provides rapid acceleration (low system time constant) and accurate positioning under heavy loading conditions.

4.4.1 The Hydraulic Power Supply

The power source for hydraulic actuation systems is high pressure oil supplied by an oil pump. Three different pump configurations are used for hydraulic purposes: the *gear pump*, the *vane pump*, and the *piston pump*. A typical configuration for each type is illustrated in Figure 4.13.

The gear pump (Figure 4.13*a*) produces a pressure increase by forcing oil through a set of rotating gears. The meshing gear teeth provide a seal to retard back flow to the inlet side. Because of the leakage inherent in its design, the gear pump is most efficient at low speeds and pressures.

The vane pump, illustrated in Figure 4.13*b*, uses an eccentric rotor to force oil to the outlet side. The net flow depends on the eccentricity, *e*, of the rotor.

The radial piston pump (Figure 4.13*c*) also uses the principle of eccentric rotation to develop a pressure differential. As a rotating piston moves in a counterclockwise direction, fluid is admitted as the stroke increases and is forced out at high pressure as the stroke decreases.

The power output for any of the above pumps can be expressed as

$$\text{hp} = 6.9 \times 10^{-3} Q \cdot \Delta P \tag{4.6}$$

where Q is the volume flow rate in m^3/sec, ΔP is the pressure differential in Pa.

In hydraulic actuation systems the oil undergoes a considerable temperature increase. By maintaining a relatively stable oil temperature, the specified properties of the hydraulic fluid are maintained. Therefore, a necessary part of all high pressure hydraulic networks is an oil cooling device.

4.4.2 Hydraulic Actuators

The two most common hydraulic actuators in NC machinery are cylinders and motors. In general the hydraulic cylinder is used only for short stroke

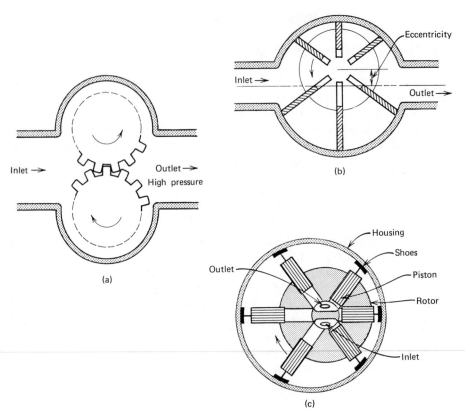

Figure 4.13 Hydraulic pumps. (a) Gear pump. (b) Vane pump. (c) Radial piston pump.

applications, i.e., 0.6 m or less. For longer travel and heavier work loads, the hydraulic motor is used.

Effectively, a hydraulic motor is a reverse pump. High pressure fluid enters the motor which extracts power from the fluid and transmits it to a rotary drive shaft.

The volume flow rate, Q_n, delivered to the motor can be expressed as the ideal flow rate, Q_i, minus flow losses due to leakage, Q_L, and compressibility, Q_c. That is

$$Q_n = Q_i - Q_L - Q_c \tag{4.7}$$

Equation (4.7) can also be written as

$$Q_n = D_m \dot{\theta}_m \tag{4.8}$$

where D_m is the volumetric displacement of the motor per radian and $\dot{\theta}_m$ is the motor speed.

Equating hydraulic power delivered to the motor, PQ_n, and mechanical power generated by the motor, $T_m \cdot \dot{\theta}_m$, we have

$$PQ_n = T_m \dot{\theta}_m \tag{4.9}$$

where P is the pressure supplied to the motor and T_m is the motor torque. Substituting equation (4.8) into (4.9), the motor torque can be expressed in terms of input pressure,

$$T_m = D_m P \tag{4.10}$$

From equations (4.7) and (4.8)

$$D_m \dot{\theta}_m = Q_i - Q_L - Q_c \tag{4.11}$$

It can be shown (Reference 3) that

$$\begin{aligned} Q_i &= K_p L \\ Q_L &= (K_L / D_m) T_m \\ Q_c &= (K_c / D_m) s T_m \end{aligned} \tag{4.12}$$

where L is the stroke length of the motor, and K_p, K_L, and K_c are system constants. Hence from equation (4.11)

$$D_m \dot{\theta}_m = K_p L - \frac{K_L}{D_m} T_m - \frac{K_c}{D_m} s \, T_m \tag{4.13}$$

The motor torque can be expressed as

$$\mathbf{T}_m = (\beta + sJ)\dot{\theta}_m + T_L \tag{4.14}$$

where β, J, and T_L are defined as for the electric motor.

From equations (4.13) and (4.14), it therefore follows that

$$\dot{\theta}_m = \frac{D_m K_p L - (K_L + sK_c)T_L}{s^2 K_c J + (K_c \beta + K_L J)s + (D_m^2 + K_L \beta)} \tag{4.15}$$

which yields a second order characteristic for the hydraulic motor.

The hydraulic motors used in NC machines have constant displacement pistons. A typical radial arrangement for a hydraulic motor is illustrated in Figure 4.14.

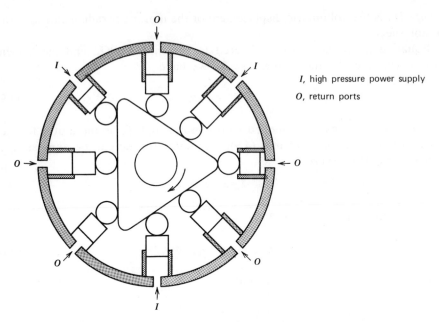

I, high pressure power supply

O, return ports

Figure 4.14 Radial hydraulic motor cross section.

4.4.3 Hydraulic Power Amplification

Power amplification in a hydraulic system requires that an electrical signal processed in the MCU be converted to a proportionate hydraulic pressure. The hydraulic fluid flow rate must be controlled so that a wide range of motor speeds is available. In numerical control machines a common hydraulic power amplification device is the *four-way proportional servo valve.*

A four-way valve amplifier contains two actuator ports which serve as the supply and return to the servomotor, one or more high pressure inlet ports from the pump, and a drain port returning fluid to the pump and cooling system. A simplified version of a four-way amplifier is illustrated in Figure 4.15.

High pressure oil enters the valve at ports A_1 and A_2. So that oil will pass from A_1 or A_2 to one of the actuator ports, C or D, the piston (solid shading) must be displaced to the right or left. If the piston is displaced to the right, high pressure oil flows from A_1 to C and returns from the motor through port D, subsequently draining through B.

The displacement of the piston can be accomplished if the pressures on faces F_1 and F_2 are unequal. To obtain this pressure difference the *flapper valve* (Figure 4.15) must be displaced so that the flow through one orifice is greater than the flow through the other. Referring to the figure inset, the torque motor

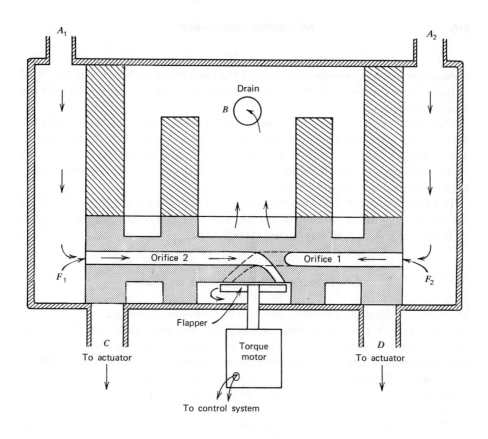

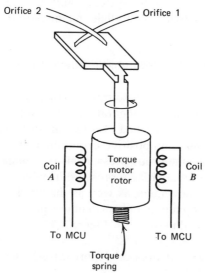

Figure 4.15 Four-way proportional servo valve and flapper control.

is designed so that the rotor turns only if the current through coil A is not equal to the current through coil B. The rotor is held in a central rest position by the torque spring.

If an imbalance in current from the control system causes the flapper valve to rotate to the right, the flow in orifice 2 is reduced while the flow through orifice 1 is increased. Hence, pressure builds on face F_1 and the piston moves to the right. This process enables a low level electrical signal to be transformed to high pressure input to the servomotor.

The four-way hydraulic amplifier with a flapper valve control is only one of many ways in which hydraulic power amplification can be accomplished. Refer to References 4, 5, and 6 for more details.

4.5 Pneumatic Actuation Systems

Pneumatic actuation systems operate on the same principle as hydraulic systems except that pneumatic devices use compressed air as the power transmission medium. NC equipment using pneumatic actuation devices is characterized by somewhat slower system response than hydraulically driven equipment.* The advantages of pneumatic devices lie in the readily available working medium—air—and in the extremely high pressures that can be accommodated.

The principles of operation and analysis of pneumatic actuation devices are quite similar to those of hydraulic systems, and it is not uncommon for the same basic components (with only slight modification) to be used in both systems.

4.5.1 The Pneumatic Power Supply

The air *compressor* is the power source for pneumatic actuation systems. Although many different pump configurations are available, the most common compressor designs can be classified into two groups—*positive displacement* compressors and *dynamic* (nonpositive displacement) compressors.

In the positive displacement compressor, air is confined in a progressively diminishing space, thereby creating a pressure increase. The compressor uses reciprocating or rotary movement as well as diaphragm action to cause this pressure rise. The most common and efficient arrangement is the reciprocating piston within a cylinder, illustrated in Figure 4.16a.

The dynamic compressor imparts a high velocity to the gas. A typical example is the centrifugal compressor (Figure 4.16b) which uses an impeller to

* Response is slowed because the air will compress significantly before activation begins.

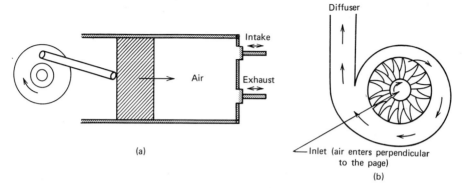

Figure 4.16 Pneumatic power supplies. (a) Positive displacement compressor, piston cylinder arrangement. (b) Centrifugal compressor.

impart considerable kinetic energy to the air as it is *turned* from an axial to a radial direction. Static pressure is built up as the air moves through the diffuser, causing the pressure increase.

For extremely high pressure air, multistage compressors are used to boost pressure in stages. Most high pressure pneumatic systems make use of filters to remove dust from the air and dehydrators to remove moisture.

4.5.2 Pneumatic Actuators

The most common pneumatic actuator in NC systems is the double-acting pneumatic cylinder, but this cylinder-piston arrangement makes possible only a limited traverse motion. Another actuator is the pneumatic motor. Unlike its hydraulic counterpart, the torque developed by a pneumatic motor is proportional to the supply pressure and independent of motor speed.

4.5.3 Pneumatic Power Amplification

With modifications, the four-way proportional valve can be used in pneumatic actuation systems. A typical configuration set up to drive a cylinder is illustrated schematically in Figure 4.17. In this case the flapper control discussed in the previous section has been replaced by a controller that moves the piston directly.

We can examine the transfer function for this system by considering how a control movement, x, will affect the output velocity, $\dot{y}$. To simplify our discussion we assume that pressure variations are small and compressibility

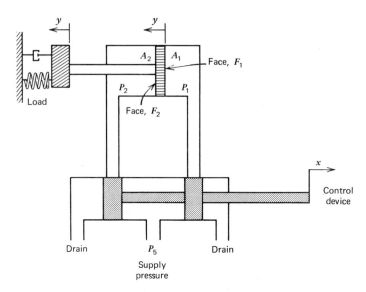

Figure 4.17 Four-way valve-actuator schematic.

effects are negligible. Then, the flow through the supply port to the face F_1 of the cylinder can be shown to be

$$Q_1 = K_d x \sqrt{P_s - P_1} \qquad x > 0 \qquad (4.16a)$$

where K_d is a constant which depends on the part geometry, and fluid density and properties. Similarly, the flow to face F_2 can be written

$$Q_2 = K_d x \sqrt{P_2} \qquad x > 0 \qquad (4.16b)$$

For negative control displacement (i.e. $x < 0$)

$$Q_1 = K_d x \sqrt{P_1} \qquad (4.17a)$$

$$Q_2 = K_d x \sqrt{P_s - P_1} \qquad (4.17b)$$

Now, recalling the velocity $\dot{y} = Q/A$, the preceding expressions become

$$\dot{y} = \frac{Q_1}{A_1} = \frac{K_d x \sqrt{P_s - P_1}}{A_1} \qquad x > 0 \qquad (4.18a)$$

$$\dot{y} = \frac{Q_2}{A_2} = \frac{K_d x \sqrt{P_2}}{A_2} \qquad x > 0 \qquad (4.18b)$$

and

$$\dot{y} = \frac{Q_1}{A_1} = \frac{K_d x \sqrt{P_1}}{A_1} \qquad x < 0 \tag{4.19a}$$

$$\dot{y} = \frac{Q_2}{A_2} = \frac{K_d x \sqrt{P_s - P_2}}{A_2} \qquad x < 0 \tag{4.19b}$$

The piston transmits a force, F, which is the resultant of the forces acting on the faces. Hence

$$F = P_1 A_1 - P_2 A_2 \tag{4.20}$$

Solving equations (4.18) and (4.19) to eliminate P_1 and P_2 and using equation (4.20), we get

$$\dot{y} = K_d \sqrt{\frac{P_s A_1 - F}{A_1^3 + A_2^3}} x \qquad x > 0 \tag{4.21a}$$

$$\dot{y} = K_d \sqrt{\frac{P_s A_2 + F}{A_1^3 + A_2^3}} x \qquad x < 0 \tag{4.21b}$$

Thus, the velocity response of the piston is given in terms of a transfer function, and because we assumed negligible compressibility, the above analysis is equally valid for a hydraulic system.

4.6 A Comparison of Actuation Systems

Continuing improvements in actuating system components would make a detailed discussion of specific device characteristics obsolete. The interested reader is therefore advised to contact control and machine tool manufacturers for the latest component specifications.* Some general areas for comparison include:

1. *Power Consumption.* Electromechanical actuation systems draw power only when the machine member is in motion. Hydraulic and pneumatic systems must have a high pressure power supply available at all times; hence, pumps run continuously.
2. *Size.* Because hydraulic and pneumatic systems require pumps and compressors as well as high pressure supply lines, they require more space than electromechanical systems of equal power rating.

---

* *The NC/CAM Guidebook and Directory* (Reference 7) published yearly, provides the address of every major NC related manufacturer and supplier.

3. *Response.* The bandwidth of a hydraulic system is larger than that for electromechanical or pneumatic actuation. Hence, the hydraulic system can respond to more commands per second than its counterparts. The response time for electromechanical and pneumatic systems are approximately equivalent.
4. *Power.* High pressure systems are capable of providing a higher continuous power output than electromechanical systems.
5. *Environment.* High pressure systems generally require filtering, cooling, and/or dehydrating apparatus to maintain proper transmitting media characteristics. Hydraulic and pneumatic valves become easily clogged with dirt. The major consideration for electromechanical systems is temperature. The operation of dc motors must take place within their thermal range.
6. *Application.* In general, hydraulic actuation systems are used for multiaxis contouring applications where a rapid and smooth response under conditions of transient or heavy loading is required. Pneumatic and electromechanical actuation is most often used in positioning systems where slower response can be tolerated.

4.7 Basic NC Electronics

The control functions for the logic operations performed by the MCU depend upon electronic circuitry. These operations are carried out by circuitry that is continuously developing. We therefore consider only the basic circuit components: *AND and OR gates, flip-flops, memory* circuits, and *counters.*

The MCU is a special purpose computer, and computer terminology and logical symbols are used to describe component circuitry. The logic operations performed by the MCU are processed by a *gate* which is either open or closed to the transmission of signals or pulses. These gate inputs and outputs are *true* values (logical *1*) and *false* values (logical *0*).

4.7.1 AND and OR Gates

The circuits illustrated in Figure 4.18 are simple analogies to AND and OR gates. In Figure 4.18*a* switches S_1 *and* S_2 must be closed for an output signal to be present across the resistor, *R*. If we assign logical *1* to the closed condition and *0* to the open condition, all possible logical combinations can be illustrated, as shown in Figure 4.20*a*. This *truth table* indicates the combination of logical inputs that will cause a logical output of *1* (Reference 8).

Figure 4.18*b* illustrates a simple OR gate circuit. It is evident that if either switch S_1 *or* S_2 is closed, a signal will pass.

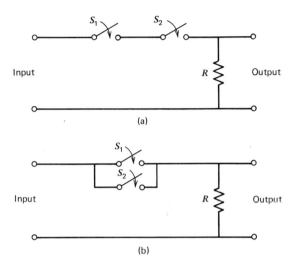

Figure 4.18 (a) AND and (b) OR gate circuits.

To streamline the specification of logic circuits, special symbols have been developed to indicate gate circuits. Symbols for AND and OR circuits are illustrated in Figure 4.19a,b.* The AND and OR circuits used in the MCU make use of solid state electronic components—*not* mechanical switches.

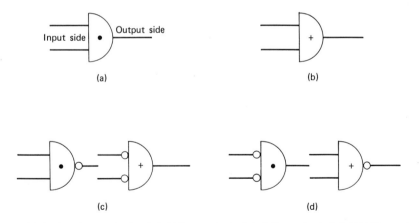

Figure 4.19 Logic symbols. (a) AND symbol. (b) OR symbol. (c) NAND symbols. (d) NOR symbols.

* It should be noted that slight variations in symbols are found in different texts.

4.7.2 NAND and NOR Gates

NAND (an acronym for *not and*) and NOR (*not or*) gates produce output opposite to AND and OR gates, for the same input. The truth table illustrated for the NAND circuit in Figure 4.20b corresponds to this rule. NOR will produce a *1* output only when both inputs are *0*, whereas NAND will produce a *1* output for all situations *except* when both inputs are *1*. The logical symbols for NAND and NOR are shown in Figure 4.19c, d. An additional feature is that OR and AND produce output pulses of the same polarity (Figure 4.18) as input, whereas NAND and NOR devices produce output pulses of opposite polarity.

Consider the circuit diagram for a typical gate shown in Figure 4.21. For the PNP type transistor* shown in Figure 4.21, a negative pulse in the base-emitter

S_1	S_2	Output
1	1	1
0	1	0
1	0	0
0	0	0

(a)

S_1	S_2	Output
1	1	0
0	1	1
1	0	1
0	0	1

(b)

Figure 4.20   (a) AND and (b) NAND (not AND) truth tables.

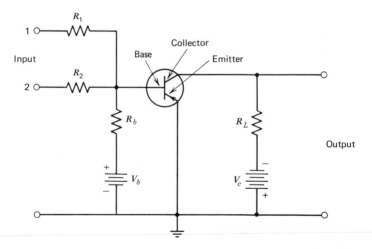

Figure 4.21   A gate circuit for NOR or NAND.

* Readers who are unfamiliar with transistor operating characteristics should consult a textbook on electrical circuit theory (e.g., Reference 9).

circuit increases current flow through the emitter-collector circuit. This has the effect of producing a *positive going* output voltage across R_L. In the circuit shown, a positive *bias voltage*, V_b, effectively cuts off current flow in the emitter-collector circuit until a negative input pulse of sufficient magnitude drives the transistor to conduction.

A *negative going* input pulse through R_1 or R_2 produces a positive going output pulse across R_L. Hence, by definition the circuit is a NOR gate. However, we shall see that under different circumstances, the circuit may also be used as a NAND gate.

A negative going output pulse is generated due to the voltage, V_c, if the transistor is *cut off*. When positive going pulses are applied at 1 and 2 such that the base is at least as positive as the emitter, the transistor cuts off and a negative going output pulse results. Under these conditions the circuit is a NAND gate. By choosing the appropriate values for bias and input voltages, the same circuit can therefore act as either a NOR or a NAND gate.

4.7.3 The Flip-Flop Circuit

As illustrated in Figure 4.22*a*, two NOR gates can be *cross-coupled* to form a simple memory circuit called a *flip-flop*. The flip-flop is the basic logic circuit used in pulse generators, buffer memory, and binary counters. When the double switch is in the OFF position, the flip-flop will *remember* whether the switch was last in the SET or RESET position. For example, with the switch in the SET position, a *1* enters gate 1 at *A*. Recalling that a NOR gate will output *0* for one or more *1* inputs, it can be seen that *0* is applied to *D* at gate 2. Because *C* is also *0* (note the switch position), gate 2 outputs a *1* to *B*. When the switch is

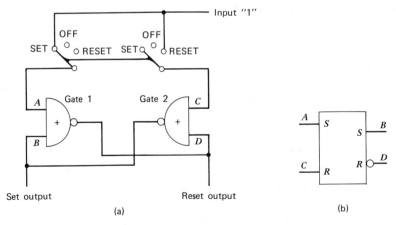

Figure 4.22 (a) Flip-flop schematic, and (*b*) logic symbol.

moved to OFF, gate 1 will continue to output *0* by virtue of the *1* at *B* (*A* has a value of *0* at OFF). Hence, the flip-flop *remembers* that a *1* was applied to *A* by maintaining a SET output of *1* and a RESET output of *0*. Similarly, it can be shown that if the SET output is *0*, and the RESET output is *1*, a *1* was last applied at *C*.

The logic symbol for a simple flip-flop is illustrated in Figure 4.22*b*. The inputs are on the left and the outputs on the right. Correspondence to the original circuit (Figure 4.22*a*) is indicated. The presence of the small circle indicates that *0* will be output when the flip-flop is in the SET position. Absence of the circle indicates a *1* output.

4.7.4 Pulse Generation

The flip-flop cannot be used if inputs are simultaneous because a *1* simultaneously applied at *S* and *R* produces an indeterminate output state. Simultaneous input can be accommodated if the flip-flop is *self-steered*; that is, if the flip-flop is in a SET state the input is steered so that it can only be reset and vice versa.

Self-steering is implemented by gating the input as illustrated in Figure 4.23. By definition, when the flip-flop is set, the input at *D* is *0* and at *C* is *1*. Hence, regardless of the *A* input, the output of gate 1 must be a *0* which will not affect the flip-flop state. When input *B* goes to *0*, *G* is *1* and the flip-flop resets. The input is steered so that *E* and *G* can never be *1* simultaneously.

The self-steering mechanism can be used to produce a signal oscillator which makes a square wave pulse generator. The flip-flop oscillates continuously between SET and RESET if the inputs *A* and *B* (Figure 4.23) are maintained at *0*. By replacing a filter circuit before the input gates, the frequency of oscillation can be determined.

In many MCU circuit components, a more complicated flip-flop, which can

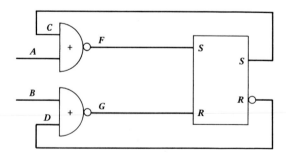

Figure 4.23   Self steering flip-flop circuit.

be set or reset by a single logical input, is used. Such devices are utilized in both memory components and binary counters.

4.7.5 Buffer Memory

Data that is placed in the buffer storage of an NC machine control unit is stored in binary form, that is, by a combination of binary digits (*bits*). As we have seen, the flip-flop is capable of representing two states: *0*, or the RESET state, and *1*, the SET state. Hence, a binary memory circuit is constructed using a series of flip-flops.

Other criteria must also be satisfied by a memory circuit. It should provide for erasure of all data and be updatable. Data within the memory must be read without erasure. The circuit illustrated in Figure 4.24 represents a memory element which can store the binary numbers 0000 to 1111 (0 to 15, decimal). Individual inputs set the required bits. For example, if flip-flops I and III are set, the binary number 0101 (5, decimal) is indicated. Further information on the characteristics of these devices is presented by Ertell (Reference 10).

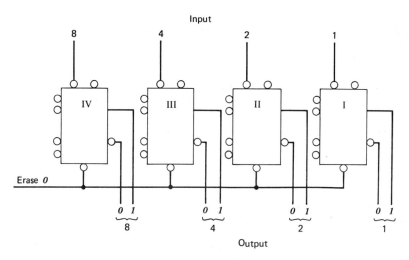

Figure 4.24 A simplified memory element. (From Ertell, G., *Numerical Control*, copyright © 1969 by John Wiley & Sons. Reproduced with permission.)

4.7.6 Binary Counting Circuits

Many comparison and control functions within the MCU require circuits with the ability to perform addition and subtraction. The *binary up counter* and the *binary down counter* are used to implement these simple arithmetic procedures.

Binary counting circuits are comprised of flip-flops whose various inputs are gated so that sequentially increasing (*up*) or decreasing (*down*) combinations of states will be generated. Counting circuits are generally driven by a stream of clock pulses.

Addition can be performed with one up counter and one down counter. The up counter is initialized to contain the augend and the down counter is preset with the addent. As clock pulses are simultaneously applied to each circuit the up counter will increase one unit for each clock pulse, while the down counter decreases one unit. When the down counter reaches zero, the value in the up counter must be the sum of the augend and addend.

Subtraction is performed using two down counters. The counters are initialized with the subtrahend and minuend, and simultaneous clock pulses are applied until the counter preset with the minuend contains a zero value. At this point the other counter contains the result of the subtraction.

References

1. Lloyd, T. C., *Electric Motors and Their Applications*, Wiley-Interscience, New York, 1969.
2. ———. *Electric Motors and Controls, a Machine Design*, Reference issue, vol. 47, April 24, 1975.
3. Raven, F. H., *Automatic Control Engineering*, McGraw-Hill, New York, 1968.
4. ———. *Fluid Power, a Machine Design*, Reference issue, vol. 46, September 1974.
5. Shearer, J. L., "Conversion, Transmission and Control of Fluid Power," *Handbook of Fluid Dynamics*, V. L. Streeter, ed., McGraw-Hill, New York, 1961.
6. Pippenger, J. J., and Hicks, T. G., *Industrial Hydraulics*, 2nd ed., McGraw-Hill, New York, 1970.
7. *NC/CAM Guidebook and Directory*, a special issue of *Modern Machine Shop*, Gardiner Publications, Cincinnati.
8. Booth, T. L., *Digital Networks and Computer Systems*, Wiley, New York, 1971.
9. Smith, R. J., *Circuits, Devices and Systems*, 2nd ed., Wiley, New York, 1966.
10. Ertell, G. G., *Numerical Control*, Wiley-Interscience, New York, 1969.

Additional References

Childs, J. J., *Principles of Numerical Control*, The Industrial Press, New York, 1965, pp. 163–81.

Walker, J. R., "Types and Selection of Positioning Systems," *Numerical Control in Manufacturing*, F. W. Wilson, ed., ASTME, 1963, pp. 132–47.

Rexrode, L. O., "Resolvers vs. Encoders for NC," *Control Engineering*, vol. 19, April 1972.

Bailey, S. J., "Stepper Motors Respond to Direct Digital Command," *Control Engineering*, vol. 21, January 1974, pp. 46–48.
Jackson, J. F., and Bell, R., "An Electrohydraulic Stepping Motor for Numerically Controlled Machine Tools, *M.T.D.R. Proceedings*, vol. A, September 1970.

Problems

1. A small NC positioning table has a maximum travel of 0.15 m along the x-axis. Full travel is to be measured by a single rotary transducer using the geared connection illustrated in Figure 4.2c. Develop a specification for the number of gear teeth and gear diameters required for each gear in the transmission.

2. Windowing is one method that eliminates ambiguous output from an encoder. Discuss other methods which might be used as antiambiguity techniques.

3. One complete revolution of a pulse generating encoder attached to an NC positioning table represents 0.25 m of motion. If the encoder contains a rotating disk with 2500 etched lines, to what linear distance does each pulse correspond?

4. Using simple graphs of signal vs time, discuss transducer signal processing for a rotary resolver. The signals for all components in Figure 4.5 should be represented.

5. A stepping motor contains three disks each containing 36 poles corresponding to 36 stator poles. If each of the rotor-stator assemblies is offset from the preceding assembly by 2.0°, what will be the angular displacement of the motor shaft (for each pulse), when pulses are applied sequentially to like poles on each stator?

6. Develop an electrohydraulic power amplification system for an open loop NC machine that uses a stepping motor to control the flow of high pressure oil. Provide working drawings for your design.

7. (a) A stepping motor is capable of accepting a maximum command rate of 250 pulse/sec. If each motor step is equivalent to 0.02 mm of linear table movement, what is the maximum feedrate (mpm) that this motor is capable of generating?
   (b) How many command pulses per second would be required for a feedrate of 2 mpm? Examine available literature to determine whether a stepping motor with this capacity exists.

8. (a) Given the dc motor transfer function, equation (4.5) in the text, how will changes in the motor constant, k_m, affect the speed output?
   (b) Express the speed, $\dot{\theta}$, as a function of time and E_f.

9. Signal rectification can be accomplished using a simple diode circuit. Why is it necessary to use a controlled rectifier (SCR) in a power amplification circuit?

10. A hydraulic actuation system has the following characteristics: Pump power, 750 W; operating pressure (gage), 6.9×10^5 Pa; and actuation motor speed, 600 rpm.
   (a) What is the approximate actuation motor output torque assuming no losses?
   (b) What is the volumetric displacement of the motor?
   (c) Reevaluate (a) and (b) assuming losses of 10 percent.

11. Express the time constant for the hydraulic motor in terms of system constants.

12. An NC machine uses a hydraulic cylinder to position a movable table along the vertical axis. If the cylinder area is 0.004 m², what is the minimum pump power needed to move a 1000-kg workpiece? Volumetric flow rate is 8.0×10^{-4} m³/sec.

13. Explain how a specified dead-zone can be built into a four-way proportional valve.

14. For the four-way proportional pneumatic valve (Figure 4.17), develop an expression for cylinder displacement as a function of control device displacement, x.

15. Develop a truth table for the OR and NOR gates. Given two NOR gates in series as shown in Figure P.4.1, develop a truth table for A, B, C, D.

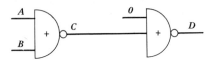

Figure P.4.1   Problem 4.15.

16. Given the gating circuit illustrated in Figure P.4.2, specify the logical value of points A through G.

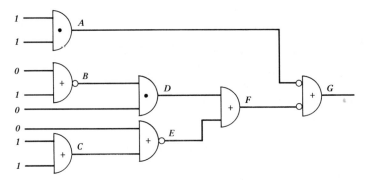

Figure P.4.2   Problem 4.16.

17. If the output of the gating circuit in Problem 16 is used as simultaneous input to a self-steering flip-flop, what state will the flip-flop be in after input?

18. Buffer storage circuits similar to that in Figure 4.24 are used to store the decimal representation of input data. This is accomplished by assigning the first circuit unit storage, the second tens storage, and so on. Indicate the state of each flip-flop in the storage array required to store the decimal number 1076. How many flip-flops would be required to store the binary representation of the number?

19. The numbers 47,032 and 36,704 are to be added using a counting circuit. The number 14,126 is to be subtracted from the result. If a single clock pulse is equivalent to 1 unit, how many pulses will be required to perform the above operations?

Chapter Five
Design Considerations for NC Machine Tools

The design of NC machine tools requires a multifaceted approach which considers both control system response characteristics and the mechanical characteristics of the machine structural members. Because the NC machine is a complex system, an integrated approach must be taken and an individual component designed so that it will conform to the criteria defined for the entire system. The most accurate control system is of limited value if the mechanical actuation components do not adequately respond to commands.

Once the performance requirements of the machine tool are determined, the overall NC system can be designed. General parameters, such as: machine function; workpiece size, type and material; accuracy and surface finish; power consumption; control features; and auxiliary functions, all lead to the proper selection of an integrated control system and machine configurations.

5.1 Design Differences between Conventional and NC Machines

The benefits of an NC machine are directly related to its accuracy, speed, and *automaticity*. However, it is these same characteristics that create design problems. A manually operated machine tool has an intelligent source for error compensation. That is, a machinist who has worked with a particular machine over a long period of time learns its characteristics. Under a given set of operating conditions, experience has taught him that one or two thousandths error results from machine member deflection. The knowledgeable operator can then compensate.

The NC machine tool can only compensate for an error that is detected and communicated to the control unit. Structural compliance,* vibration, and other

---

* The term *compliance* is used to specify elastic deformation, that is, deflection, strain, and compression due to torque and other forces.

aspects of mechanical design cannot be easily or economically monitored. For this reason, an NC machine is designed to be stronger, stiffer, and to perform to a more accurate standard than its conventional counterpart.

NC and conventional equipment achieve different dynamic performance. The NC machine elements must be capable of performing accurately under high accelerations and decelerations. Because leadscrews and gearing are an integral part of the closed loop system, mechanical inaccuracy such as backlash, windup, and deflection, which tend to produce instability, must be reduced to levels not required for manually operated machines.

5.1.1 Control System Design Considerations

The control system of an NC machine tool is required to respond accurately under both steady state and dynamic operating conditions. In previous chapters we have seen that the loop gain, K_L, and the system bandwidth, B, are two major factors that govern servo response.

The values of K_L and B have a direct influence on the *frequency response* and the *stiffness* of the servo. Frequency response is extremely important when considering system stability, whereas stiffness is directly related to the dynamic accuracy of the machine.

5.1.2 Mechanical Design Considerations

A general criterion for the design of machine structures for numerical control is to provide adequate static stiffness with the best stiffness to weight ratio for a broad range of loading conditions (Reference 1). The dynamic response of the machine is directly related to the stiffness to weight ratio. Stiff machines with low mass have a rapid dynamic response which is essential in some positioning and all contouring NC machines.

Other aspects related to the mechanical nature of the NC machine (e.g., friction, backlash, slip-stick) can adversely affect the machine performance if they are not compensated or minimized. Vibration, resulting from tool *chatter*, can have a devastating effect on the quality of surface finish and accuracy. In extreme conditions it may lead to machine tool damage.

5.1.3 Overall Design Criteria

The overall design of an NC machine tool system can be reduced to a number of interrelated steps. These include:

1. Development of specifications for steady state and dynamic operation under servo control.
2. Specification of individual component characteristics in terms of "lower-bound design criteria" (Reference 2).
3. Evaluation of operating conditions and cost based on practical application of the system.

Some additional parameters that should be considered during the evaluation of these steps are listed in Table 5.1.

As is the case with nearly all commercial engineering designs, the final NC system is developed through necessary and acceptable compromises, and cost-performance trade-offs. The best overall design will maintain acceptable system response characteristics for a wide range of variation in critical operating parameters.

Table 5.1
NC Machine Tool Design Criteria

Machine response characteristic	Component characteristics	Operating and cost considerations
Type of command signal	Undamped natural frequency	Reliability
Input configuration	Power requirement	Maintainability
Maximum feed rate	Friction characteristics	Cost of operation
Static accuracy	Inertia	Capital investment
Dynamic accuracy	Stiffness	Installation requirements
Magnitude of load	Amount of backlash	
Range of travel	Speed range	
Weight of moving members	Bandwidth	
Power source		

5.2 Lost Motion in the NC System

The term *lost motion* is used to indicate a condition in which the command position is not reached. This may be due to *backlash* in the drive system gearing, *windup* of drive shafts, or *deflection* of machine tool members. In practice the total lost motion is a combination of these factors.

The effect this has on accuracy depends on the type of control system and servomechanism that is used. Continuous path (contouring) control systems are most sensitive to inaccuracies along the tool path. Even small amounts of lost motion in each segment of motion tend to accumulate as direction or velocity is changed. For this reason major attempts are made to reduce the magnitude of lost motion.

Positioning systems are less sensitive to lost motion. Because path error need not be considered, methods for backlash compensation (see Chapter Three) are used to eliminate error due to backlash. Since static position accuracy is all that is required in PTP systems, windup and deflection are also of reduced magnitude.

To illustrate the role lost motion plays in the causes of position inaccuracy, consider the following example. If the loading on a positioning table is such that the servomotor torque produces 1° of windup in a 20-mm pitch leadscrew, lost motion can be shown to be

$$\text{lost motion} = \frac{(\text{windup}) \times (\text{pitch})}{360°}$$

$$= \frac{(1°) \times (20 \text{ mm})}{360°}$$

$$= 0.055 \text{ mm}$$

This magnitude of lost motion in the table slide movement would be considered unacceptable for some NC applications.

The effect of windup can be reduced significantly if the drive shaft transmits power to the leadscrew through gearing. For example, a 10:1 gear ratio would result in only 0.0055 mm of lost motion, one-tenth of the previous value.

Although excessive backlash is detrimental to positioning accuracy, drive gearing must have a small amount of backlash so that the gears will operate smoothly at low torque. Small amounts of backlash do not appreciably affect positioning accuracy. As an example, consider a 500-mm circumference gear powering a 20-mm pitch leadscrew. If backlash is 0.04 mm, it can be shown that only 0.002-mm lost motion is evident at the slide (Reference 3).

This demonstrates that lost motion at the high end of the gear train has less effect on position accuracy. Conversely, at the low speed end of gearing, lost motion must be carefully controlled. Lost motion can also occur in the feedback loop sensory components. For example, a rotary transducer connected to the actuated components through gearing must be configured so that most motion is virtually eliminated.

Acceptable levels of lost motion in NC equipment depend upon the size and application of the machine tool and the accuracy for which it was designed. Generally, contouring equipment has total lost motion less than or equal to 0.004 mm. Other NC machines requiring less accuracy, and most PTP equipment allows 0.01 mm to 0.02 mm lost motion.

5.3 Sources of Lost Motion

5.3.1 Backlash

Although we have discussed backlash as a phenomenon associated with gearing, other forms of mechanical backlash are also found in machine structures. Looseness in bearings and bearing mounts, play between leadscrew and nut, and the effect of cocking of the machine slide are also classified as backlash.

Rigid mounting techniques for bearings and screws eliminate one form of backlash that is associated with mechanical components. In many cases, preloading components also serves to reduce backlash.

Cocking of the machine slide is due to slight errors in alignment between the machine slide and its ways. If misalignment is severe, a point on the slide may actually move backward for a short distance before motion in the command direction commences.

A *backlash take-up* feature can be provided in the control system electronics. If the amount of backlash is known and constant, the servo response curve (Figure 3.9) can be shifted against the direction of backlash. In this way the control system will command a small opposite movement when the machine reaches the indicated zero position, thereby compensating for backlash.

5.3.2 Windup in Machine Components

Windup is defined as an angular deflection, or twisting, resulting from an applied torque to rotary members such as drive shafts and screws. The windup of a rotary component is a function of the stiffness of the member and the load to which it is subjected.

Windup in a circular shaft is illustrated in Figure 5.1. From elementary stress analysis, the shearing strain, γ, of an element on the shaft circumference is expressed as

$$\gamma = \frac{r\theta}{l} \tag{5.1}$$

where θ is the angular displacement (windup) between adjacent cross sections, r is the radius of the shaft, and l is the distance along which torque is applied. The stress, τ, on the circumference of a circular shaft due to an applied torque, T, is

$$\tau = \frac{Tr}{J} \tag{5.2}$$

where r is the shaft radius, and J is the corresponding polar moment of inertia.

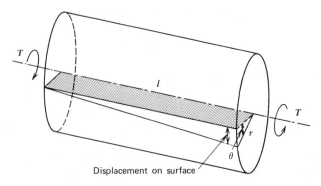

Displacement on surface

Figure 5.1 Windup due to the torque T in a shaft with circular cross section.

Hooke's law states that shearing stress, τ, is proportional to shearing strain γ,

$$\tau = G\gamma \qquad (5.3)$$

where G is the shearing modulus of rigidity.

Hence, substituting equations (5.1) and (5.2) into (5.3)

$$\theta = \frac{Tl}{GJ}$$

For a circular shaft with diameter d, $J = \pi d^4/32$, therefore, the windup, θ, can be expressed as

$$\theta = \frac{32Tl}{\pi d^4 G} \qquad (5.4)$$

5.3.3 An Example To Illustrate Windup and Lost Motion Computation

The load on the positioning table of an NC machine tool requires that the servomotor deliver 200 N · m of torque. If the motor is directly connected to a 0.5 m long leadscrew, and the leadscrew pitch is 10 mm/revolution, we can compute (a) the windup and (b) the lost motion. The leadscrew diameter is 25 mm, and $G = 7.6 \times 10^{10}$ Pa.

(a) Windup can be expressed from equation (5.4) as

$$\theta = \frac{32Tl}{\pi d^4 G}$$

$$\theta = \frac{32(200)(0.5)}{\pi (0.025)^4 (7.6 \times 10^{10})} = 0.0343 \text{ rad}$$

$$\theta = 1.965°$$

where θ is the amount of windup in the leadscrew.

(b) Lost motion can be expressed as

$$\text{lost motion} = \frac{\theta \times \text{pitch}}{360°}$$

$$= \frac{1.965 \times 0.010}{360}$$

$$= 5.46 \times 10^{-5} \text{ m of lost motion}$$

For NC machines with an accuracy 0.002 mm, the computed lost motion is not within tolerance.

5.3.4 Deflection

Deflection, as it relates to lost motion,* is produced by axial compression of the leadscrew as it transmits the necessary force to drive a positioning table. In hydraulic actuation systems it is associated with compression of the piston rod.

Compression of a circular cross-section member is illustrated in Figure 5.2. From elementary stress analysis, strain ϵ is defined as

$$\epsilon = \delta/l \tag{5.5}$$

where δ is the total compression (or elongation) of the member. The normal stress σ is expressed as

$$\sigma = F/A = E\epsilon \tag{5.6}$$

where F is the axial force, A the component area, E the modulus of elasticity, and ϵ the normal strain. Using equations (5.5) and (5.6), the longitudinal compression for any screw or shaft of circular cross section is

$$\delta = \frac{4Fl}{\pi d^2 E} \tag{5.7}$$

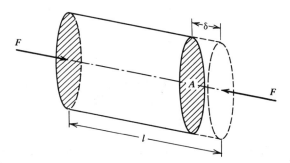

Figure 5.2   Compression due to the force F applied along the axis of a shaft.

* Other NC machine members are subjected to bending resulting from cutting forces and dynamic loads.

To minimize compression the most significant parameter, the member diameter, should be increased (equation 5.7). For this reason, drive components in NC machines generally have large diameters.

Drive component deflection due to bending occurs in the pinion drive shaft for systems using rack and pinion drive. Bending results from forces transmitted to the pinion gear through the shaft.

In an earlier discussion of hydraulic actuation systems, the compressibility of the hydraulic fluid was neglected to simplify the mathematics. However, a spring constant, which is due to fluid compressibility, can be associated with a double-acting cylinder (Figure 4.17). It can be shown that the minimum spring constant (stiffness) for a double-acting cylinder is

$$k_{cyl} = \frac{2\pi^2 B d^4}{l} \tag{5.8}$$

where B is the oil bulk modulus, d is the cylinder diameter, and l is the cylinder length.

Equation (5.8) shows that to maintain the same drive stiffness, the piston area squared must be increased in direct proportion to the stroke length. Hydraulic fluid compressibility significantly affects system response when the volume of oil between the servo valve and the piston is large.

5.3.5 An Example To Demonstrate the Effect of Stiffness on Lost Motion

An NC positioning machine uses a hydraulically driven piston to move a 2250-N load in the vertical direction. If frictional forces are neglected, what is the minimum-diameter piston for which lost motion will be less than 0.0002 mm? The piston is to be machined from stainless steel ($E = 1.9 \times 10^{11}$ Pa) and is 0.5 m long.

From equation (5.7), diameter can be expressed in terms of other parameters as

$$d = \left[\frac{4Fl}{\pi E \delta} \right]^{1/2}$$

$$= \left[\frac{4(2250)(0.5)}{\pi (1.9 \times 10^{11})(2 \times 10^{-7})} \right]^{1/2}$$

$$= 0.194 \text{ m}$$

the diameter required to maintain lost motion at less than 0.0002 mm.

Since even extremely accurate servo positioning is normally 0.002 mm, the above component would seem to be grossly *overdesigned*. Let us recompute the diameter for an acceptable lost motion of 0.002 mm:

$$d = \left[\frac{4(2250)(0.5)}{\pi (1.9 \times 10^{11})(2 \times 10^{-6})} \right]^{1/2}$$

$$d = 0.061 \text{ m}$$

Hence, it is important to choose design criteria that take servo system accuracy into account. By using a more realistic accuracy requirement, we have reduced the piston diameter by over 60 percent.

5.3.6 Friction on Machine Slides

The beneficial damping effect of viscous friction was illustrated in Chapter Two. However, dry friction is not desirable and is minimized in NC machines. Dry friction leads to a phenomenon called *stick-slip*.

Under conditions of low normal stress found on NC machine slides, Amonton's law is applicable. Thus,

$$F = \mu N$$

where F is the friction force resulting from a coefficient of friction μ and normal force N.

The initial large coefficient of friction between the table and slides requires a considerable frictional force to initiate movement, that is, to overcome the *sticking* of the components. Once motion begins, μ falls and the drive force necessary to sustain movement decreases rapidly, causing the machine table to move beyond the point where driving force equals the sliding friction force. This *slipping* action results from the inertia of the table and the spring effect due to a rapid release of energy. Although the stick-slip phenomenon only occurs at low feedrates, it is detrimental to system performance and should be minimized. In addition, because the control system tends to compensate, stick-slip may cause instability of the positioning table.

Friction contributes to lost motion in an indirect manner. The dry friction force must be overcome by a large applied force, and therefore windup and deflection (both directly proportional to load) of machine drive members will also be large. When dry friction is minimized, the drive system compliance is effectively reduced.

Antifriction bearings are used to reduce both the coefficient of friction between the machine slide and ways, and the phenomenon of stick-slip. Two basic types of bearings are used—rolling bearings and hydrostatic bearings. Both types result in low coefficients of friction and have inherent advantages and disadvantages. The choice of a particular bearing depends upon normal force level, cost, and desired accuracy.

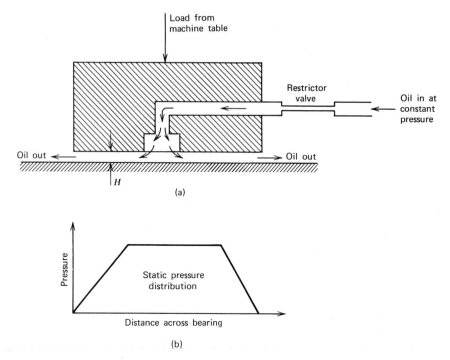

Figure 5.3 (a) Hydrostatic bearing schematic, and (b) pressure distribution.

To illustrate the use of antifriction devices in NC machine tools, consider the hydrostatic bearing schematically illustrated in Figure 5.3a. Hydraulic fluid from a constant pressure source flows through the bearing and is discharged over the bearing area in such a manner that a pressure distribution of the form shown in Figure 5.3b results. The pressure is balanced by a downward load which enables the bearing to ride on a layer of fluid, H.

The advantages of the hydrostatic bearing include the elimination of the stick-slip phenomenon and a reduction in the coefficient of friction to approximately 10^{-5}. Because the surfaces are separated, wear is negligible, and errors in the accuracy of the ways are averaged, enabling less costly machined surfaces to be used. Hydrostatic bearings are limited to low normal stress applications because of film breakdown.

5.3.7 An Example of the Stick-Slip Phenomenon

The combined weight of an NC positioning table and workpiece is 1250 N. The table is hydraulically actuated along each axis by a piston, 0.5 m long and 25 mm in diameter ($E = 1.93 \times 10^{11}$). The coefficient of static friction, μ_s, between the table and ways is 0.40. Once the table is in motion, the coefficient of kinetic friction is

$$\mu_k \approx 0.75\mu_s$$

The force required to set the block in motion, the force to keep it in motion, and the amount of lost motion which will be evident before and after motion begins can be calculated.

Recalling Amonton's equation

$$\mu = F/N$$

where F is the driving force and N is the normal force, in our examples the combined load. Hence,

$$F_s = \mu_s N = (0.40)(1250)$$

$$= 500 \ N, \text{ the force required to set the table in motion}$$

$$F_k = \mu_k N = 0.75 \ (0.40)(1250)$$

$$= 375 \ N, \text{ the force required to maintain motion}$$

To determine lost motion we use equation (5.7)

$$\delta_s = \frac{4(500)90.5)}{(0.025)^2(1.93 \times 10^{11})} = 2.63 \ \mu \, m$$

where δ_s is the compression due to *sticking* (start up) force; and

$$\delta_k = \frac{4(375)(0.5)}{(0.025)^2(1.93 \times 10^{11})} = 1.98 \ \mu \, m$$

where δ_k is the compression due to slip (kinetic) force.

Actually, lost motion would only result from compression due to δ_k. In this example the amount of lost motion is within accuracy limits for most applications.

5.3.8 An NC Design Problem Related to Cutting Forces

An NC machine is to be designed to undertake an orthogonal shaping operation using high strength work materials. Using predictive machinability information based upon the Ernst and Merchant (Reference 4) model, relate (in equation form) machining and material parameters to table piston diameter and lost motion. Table friction should not be ignored.

The NC table has to transmit two forces due to: (1) cutting and (2) friction. These result from the cutting thrust and component weight.

The machining thrust force, F_t, normal to the table is

$$F_t = \frac{F_s - F_c \cos \phi}{\sin \phi} \tag{i}$$

And

$$F_s = \frac{\tau b t_1}{\sin \phi}$$

where F_s is the primary shear force, τ primary shear stress, b width of cut, t_1 undeformed chip thickness, and ϕ primary shear plane angle (Figure 5.7).

The cutting force, F_c, in the direction of machining is given by

$$F_c = \frac{\tau b t_1 \cos(\beta - \alpha)}{\sin \phi \cos(\phi + \beta - \alpha)} \qquad \text{(ii)}$$

When the table moves, the resulting friction force, F_F, is

$$F_F = \mu F_t + \mu N$$

where μ is the effective coefficient of friction and N the weight of the component.

The total force, F, to be transmitted by the piston is

$$F = F_c + F_F$$

and from equation (5.7) the longitudinal compression, δ, is

$$\delta = \frac{4Fl}{\pi d^2 E}$$

Substituting equations (i) and (ii) into (5.7):

$$\delta = 4l \left\{ \mu \left[\tau b t_1 \left(1 - \frac{\cos(\beta - \alpha)}{\sin \phi \cos(\phi + \beta - \alpha)} \cos \phi \right) \right] + \mu N \right.$$
$$\left. + \frac{\tau b t_1 \cos(\beta - \alpha)}{\sin \phi \cos(\phi + \beta - \alpha)} \right\} \bigg/ \pi d^2 E$$

5.4 NC Machine Tool Vibration

Vibration in an NC machine tool leads to poor surface accuracy and finish, positioning inaccuracies, and the possibility of control system instability. The two basic forms of machine vibration are vibration due to resonance in the actuation system, and vibration due to combined interaction of the forces generated during the metal cutting process.

5.4.1 Machine Tool Resonance

The power regulation and mechanical members of the NC machine tool actuation system have a number of resonant frequencies. It is critically important that a resonant frequency does not lie within the operating band-width of the machine tool. For this reason, a proposed NC system must be thoroughly analyzed so that undamped natural frequencies can be predicted.

To reduce windup and deflection in machine tool members, the stiffness of each member must be sufficient to counteract applied loads. Therefore, machine members tend to have large mass and inertia. Mass not contributing to stiffness is eliminated in well designed machines.

Low mechanical resonant frequencies are caused by two major factors: lack of stiffness in the drive components and the mass of the actuated component itself. Referring to the simple spring mass and torsional analogies in Figure 5.4, it can be shown that the natural frequency of the spring-mass system is

$$f_n = \frac{1}{2\pi} \sqrt{\frac{k_s}{m}} \tag{5.9a}$$

and the natural frequency for the torsional system is

$$f_n = \frac{1}{2\pi} \sqrt{\frac{k_T}{J}} = \frac{1}{2\pi} \sqrt{\frac{\pi d^4 G}{32 lJ}} \tag{5.9b}$$

From equations (5.9) it is evident that as the mass, m, or inertia, J, increases, the natural frequency will decrease. In an NC actuation system, the stiffness is provided by the leadscrew, piston, or other drive transmission device, while the mass and inertia are provided by a combination of actuation system compo-nents.

In the analysis of a proposed design for an NC machine, the inertia of the individual system elements is reduced to one *resultant compliance* and a *resultant inertia* to be driven by the servomotor. In this way a rough estimate of the system's natural frequency is obtained.

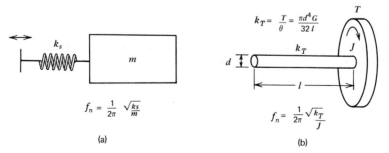

Figure 5.4 (a) Spring-mass analogy, and (b) torsional analogy used in vibration analysis.

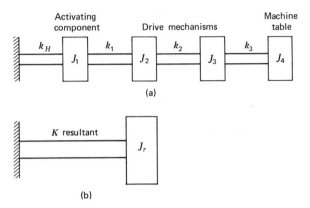

Activating
component   Drive mechanisms

Machine
table

k_H k_1 k_2 k_3

J_1 J_2 J_3 J_4

(a)

K resultant

J_r

(b)

Figure 5.5 (a) Simplified dynamic model of an NC system, and (b) the resultant system.

Figure 5.5 provides a simplified illustration of this procedure. Each element in the actuation system is represented as an inertia, and each compliance is represented as a stiffness. The multidegree of freedom system illustrated in Figure 5.5a may be reduced to the system in Figure 5.5b using Holzer's method (Reference 5).* Equation (5.9b) is then applied to find the undamped natural frequency. For a better approximation frictional damping can also be added to the model.

Low inertia is important for reasons other than frequency considerations. Low inertia systems require less servomotor power and respond to acceleration more readily. Therefore, when mass is reduced without sacrificing stiffness, system response and stability are improved.

Inertia that does not contribute to stiffness is introduced when power transmission gearing and clutches are used. Although the NC machine configuration dictates the type of power transmission, judicious application of these components will result in significant design benefits.

The presence of gearing between the servomotor and the load influences the effective inertia, coefficient of viscous friction, and torque applied to the load. Considering the geared system illustrated in Figure 5.6, it can be seen that $\theta_2 = n\theta_1$ for a gear ratio of $n:1$. Then the motor torque can be written

$$T = \tfrac{1}{2}J_L\dot{\theta}_2^2 = \tfrac{1}{2}n^2J_L\dot{\theta}_1^2 \qquad (5.10)$$

Hence, the effective inertia related to the motor is n^2J_L. Likewise, the potential energy of the leadscrew is

$$U = \tfrac{1}{2}K_2(n\theta_1)^2 = \tfrac{1}{2}(n^2K_2)\theta_1^2 \qquad (5.11)$$

* Other matrix solutions are also available.

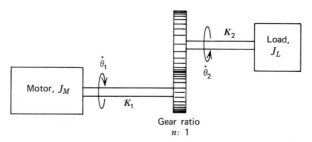

Figure 5.6   Motor and load connected to show the effect of gearing on inertia and relative stiffness.

indicating that the leadscrew has a stiffness (referred to the drive shaft) of $n^2 K_2$. In a similar manner it can be shown that the effective viscous coefficient of friction is also multiplied by n^2.

Therefore, we may state a general rule: For a geared system, the stiffness, inertia, and friction are multiplied by n^2 before they can be referenced to the drive shaft. This rule should be noted when the analysis illustrated in Figure 5.5 is performed.

5.4.2 An Example of Natural Frequency Calculation

A motor is used to drive a small rotary component with a moment of inertia of $0.05 \text{ kg} \cdot \text{m} \cdot \text{sec}^2$. The component is connected to the motor by a 16-mm-diameter steel shaft ($G = 7.6 \times 10^{10}$ Pa) which is 0.3 m long. The moment of inertia of the motor is much greater than 0.05, and the natural frequency of the system is to be determined.

Since the moment of inertia of the drive motor is much greater than the moment of inertia of the component, the motor end of the shaft can be assumed to be fixed and equation (5.9b) applies.

The stiffness in torsion, K_T, can be expressed as

$$K_T = \frac{T}{\theta} = \frac{\pi d^4 G}{32 l}$$

$$\frac{\pi (0.016)^4 (7.6 \times 10^{10})}{32(0.3)} = 1.629 \times 10^3 \text{ N} \cdot \text{m/rad}$$

Now, from equation (5.9b)

$$f_n = \frac{1}{2\pi} \sqrt{\frac{K_T}{J}}$$

$$f_n = \frac{1}{2\pi} \sqrt{\frac{1.629 \times 10^3}{(9.806)(0.05)}} = 9.17 \text{ Hz}$$

5.4.3 Machine Tool Chatter

During the metal cutting process, large forces can be generated at the tool-chip interface (Figure 5.7). These forces act on the tool and tool holder causing them to deflect. A vibratory response, called *chatter,* can result.

There are three basic types of chatter. *Arnold-type chatter* is due to a variation in force with cutting speed, causing an increase and decrease in effective work-tool velocity with change in tool deflection. The second form of chatter results from a mechanical feedback phenomenon termed *regenerative chatter. Mode-coupling* occurs when forces in one direction cause deflection in another direction (Reference 6).

As an illustration of the chatter phenomenon, we consider regenerative chatter in lathes. Metal removal is performed using single point cutting tools and the Merchant model (Figure 5.7) can be used to provide force relationship and time dependent factors.

During metal removal the tool deflects, causing the cutting edge to move away from the workpiece. This cutter movement produces a variation in the depth of cut of the machined surface and a *lump* left on the workpiece. Tool deflection, therefore, occurs one revolution (T seconds) later, when the tool

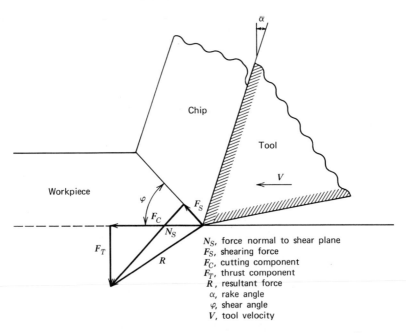

N_S, force normal to shear plane
F_S, shearing force
F_C, cutting component
F_T, thrust component
R, resultant force
α, rake angle
φ, shear angle
V, tool velocity

Figure 5.7   Merchant's cutting force model (Reference 4).

contends with the greater effective depth of cut requiring greater cutting force. Because this process is repetitious, vibration develops and instability can result.

If the workpiece is assumed to rotate with an angular velocity Ω then the force on the tool can be written as

$$f_r = k_c x + k_c e^{-(2\pi/\Omega)\omega} x \tag{5.12}$$

where the first term on the right-hand side represents force due to present position, and the second term, the force due to the increased underformed chip thickness one revolution earlier. The value k_c is termed the *cutting stiffness* and x is the direction of feed.

Figure 5.8 illustrates a block diagram of the interaction of the cutting process and machine structure which results in regenerative chatter. The cutting process block is the analog of an amplifier where gain is replaced by k_c. The machine structural response is represented by a transfer function in which k_m is the dynamic stiffness of the machine tool, and $R_m(s)$ is the dynamic response expression.

From the physical nature of the cutting process, the actual depth of cut can be expressed as

$$w(t) = w_0(t) - y(t) + \lambda y(t - T) \tag{5.13}$$

where $w_0(t)$ is the nominal cutting depth, $y(t)$ is the deflection due to forces at the current tool position, and $\lambda y(t - T)$ is the regenerative term. Recalling the use of Laplace transforms, equation (5.13) can be rewritten:

$$\mathbf{w}(s) = \mathbf{w}_0(s) - \mathbf{y}(s) + \lambda e^{-Ts} \mathbf{y}(s) \tag{5.14}$$

The coefficient λ is a nondimensional parameter representing the amount of *overlap* between successive cuts. By definition, $\lambda = 1$ for total overlap (e.g., a

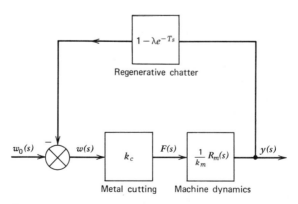

Figure 5.8 The chatter phenomenon represented in block diagram form.

cutoff operation), and $\lambda = 0$ for no overlap (e.g., a thread turning operation). For normal metal removal $0 < \lambda < 1$.

Referring to the block diagram in Figure 5.8, the transfer function can be written as

$$\frac{\mathbf{w}(s)}{\mathbf{w}_0(s)} = \left[1 + (1 - \lambda e^{-Ts}) \frac{k_c}{k_m} R_m(s) \right]^{-1} \tag{5.15}$$

The characteristic equation, which is a stability indicator, can then be expressed,

$$(1 - \lambda e^{-Ts}) \frac{k_c}{k_m} R_m(s) = -1 \tag{5.16}$$

Using the methods outlined in Chapter Two, the stability of the loop can be determined. It is interesting to examine the stability of the system as a function of the stiffness ratio, k_c / k_m, and lathe spindle speed. Such a relationship, illustrated by the graph in Figure 5.9, has been developed (Reference 7). From the graph it is evident that to obtain chatter-free operation in machining, the ratio of cutting stiffness to dynamic stiffness must be kept as low as possible.

The analysis of multipoint cutting operations such as milling is more complex. The rotation of the cutting tool causes the force vector to vary with time relative to machine structure. This leads to equations that have periodic coefficients and cannot be solved using Laplace transforms. A state variable approach (Reference 8), making use of the digital computer, is then required.

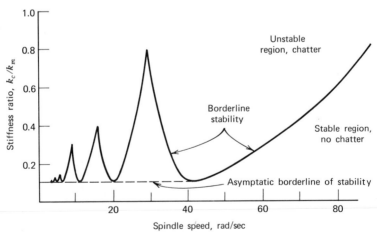

Figure 5.9 Machine stability regions as a function of stiffness and spindle speed (Reference 7).

5.4.4 Chatter Compensation

In an actual manufacturing process, many factors contribute to the chatter proneness of an NC machine tool during a particular cutting operation. Chip stiffness, k_c, is a function of workpiece material, the width of the cut, and the size, shape, and orientation of the cutting tool. The machine dynamic stiffness, k_m, is directly related to the mass and inertia of the machine structural members, size and shape of the fixturing and tooling, and, to some extent, to the operation that is being performed.

Other factors, such as size and shape of the workpiece, cutting conditions (e.g., feed, speed, depth of cut), cutting fluids, and tool sharpness, contribute to the stability of the machine tool. In summary, chatter occurs as a function of the combined elements affecting the machine system, tool, fixturing, and workpiece.

Ideally, an NC machine tool should be designed so that it is stable and chatter free over the broadest range of operating conditions. The major design factor for the elimination of chatter is keeping the stiffness ratio within the stable range (see Figure 5.9). This may require changing operational features of cutting tool geometry or width of cut to alter the value of k_c, or modifying the design of the machine tool structure to increase k_m.

It is also important that a certain amount of structural damping is built into the machine members. For stability, it can be shown* that (Reference 6)

$$\frac{2c_m\omega}{\omega_m} > \frac{k_c}{k_m} \tag{5.17}$$

where c_m is the *effective damping ratio*, ω_m is the resonant frequency of the machine tool, and ω is the angular velocity of the workpiece. The damping ratio reflects damping due to the tool frame and to the cutting process itself.

Equation (5.17) indicates that an increase in damping will reduce the chatter proneness of a machine tool. However, the expression must be considered carefully because increases in dynamic stiffness will often reduce the amount of damping.

5.5 Servomechanism Design Characteristics

A sophisticated feedback control system is applied to NC machine tools to achieve accurate positioning and velocity control without a foreknowledge of the exact magnitude and nature of the loads to be encountered. Because the NC servomechanism must respond to a wide range of disturbances, the stiffness

*Equation (5.17) applies to lathes only.

and frequency response (bandwidth) of all closed loop components are examined by the designer.

In Chapter Three we have discussed many important aspects of NC system response. In particular, the reader should review pattern errors, the system time constant, bandwidth, and cornering errors presented in Section 3.6.

5.5.1 Actuation System Response

The response of an NC actuation system is generally discussed in terms of its load handling capability and output under dynamic conditions. The *load sensitivity* of a servoactuator is defined as the force or torque output with respect to incremental changes in input (actuating) signal. The dynamic response of the servo is called *stiffness*. Both of these characteristics determine the response of an NC machine subjected to a sudden increase in loading.

The stiffness of a servo is analogous to stiffness in a spring-mass system. If the spring has high stiffness, the mass oscillates at high frequency and small amplitude; if the spring has low stiffness, low frequency oscillations of large amplitude result. To understand servo stiffness, consider the machine tool operating conditions and the typical response curve in terms of deflection illustrated in Figure 5.10.

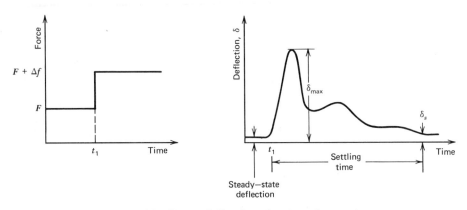

Figure 5.10 Servo deflection due to a force step.

Referring to the plot of force versus time, it can be seen that the NC machine operates under steady force conditions until the time, t_1, when an increase in loading, Δf, is encountered. The maximum transient deflection, δ_{max}, and the steady state deflection, δ_s, can be expressed (approximately) in terms of loop gain, k_L, and system bandwidth, B, in the following manner (Reference 2):

$$\delta_s = \frac{\Delta f}{(\lambda k_L)}$$

$$\delta_{max} = \frac{\Delta f}{(\lambda B)} \tag{5.18}$$

where λ is a load speed constant (N · sec/m). It is evident from equations (5.18) that large values of loop gain and bandwidth tend to minimize both transient response and steady state deflection.

5.5.2 Power Amplifier Response

The characteristics of various power amplification systems were discussed in Chapter Four. Because hydraulic systems generally have the best response characteristics, we use the valve amplifier and piston cylinder actuator as a representative example.

Oil *is* compressible. It acts as a spring attached to the piston, and when a positioning table is attached to the piston, the natural frequency of the arrangement may be computed. Recalling equation (5.8) for the stiffness, k_{cyl}, of a double-acting cylinder, and equation (5.9a) for a spring-mass system, we can write

$$f_{n,cyl} = \frac{1}{2\pi} \sqrt{k_{cyl}/M_T} = 0.707 d^2 \sqrt{\frac{B}{lM_T}} \tag{5.19}$$

where M_T is the mass of the table. Equation (5.19) indicates an important design parameter for hydraulically actuated component, that is, the product of the piston stroke, l, and the table mass, M_T. The proper choice of lM_T can be used to adjust the actuation system natural frequency.

The hydraulic amplifier (the four-way proportional valve, Figure 4.15) is subjected to varying reaction forces that are a function of oil flow rate and velocity. Under certain conditions these forces cause the spool to oscillate. The equations governing the oscillations are often nonlinear, and a more detailed discussion is not warranted in this text.*

Although hydraulic servo valve oscillation is an unstable condition and can be overcome with correct design, other factors require compensation in the servo loop. Oil temperature and pressure changes, valve hysteresis, and spool wear from abrasive particles in the oil cause the servo valve to drift. A feedback loop is designed to compensate for these factors. Generally, high gain feedback loops for position and velocity can be used to reduce the effects of servo valve drift.

---

* The interested reader should see Reference 6, Chapter Seven.

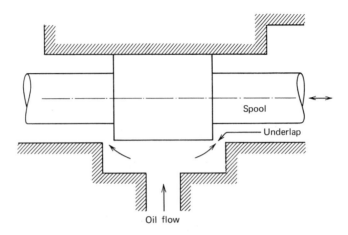

Figure 5.11   Valve underlap.

In Section 5.4 we discussed the importance of structural damping as a means of reducing machine tool chatter. When damping within the actuation system is increased, the system can tolerate a higher position loop gain, which leads to increased stiffness, improved transient response, and a reduction in path error. A number of different stabilizing methods making use of increased damping within the actuation system have been applied to the hydraulic amplifier (Reference 9); these are: (1) controlled leakage methods, (2) pressure feedback, (3) electrical compensation, and (4) hydromechanical derivative networks.

Controlled leakage methods include valve *underlap*, viscous leakage, and viscous friction. Valve underlap is illustrated in Figure 5.11. The spool is designed so that it does not fully cover the inlet and exit ports. This creates a damping (stabilizing) effect but also reduces servo stiffness and requires higher oil flow rates.

Viscous leakage allows a small amount of oil to continuously flow between the load ports. It also causes a decrease in stiffness.

Viscous friction requires no additional oil flow and does not decrease servo stiffness; however, because a large slide area is required to achieve the necessary damping effect, practical application is difficult.

5.5.3 Amplifier Stability

The gain versus frequency characteristics of signal amplifiers and the power amplifier can have a profound effect on NC machine tool system stability. A typical amplifier gain versus frequency plot is shown in Figure 5.12*a*. The gain

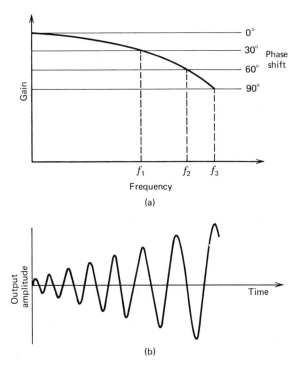

Figure 5.12 (a) Amplifier gain versus frequency. (b) Output amplitude when gain is greater than one for a 90° phase shift.

of an amplifier almost always decreases as frequency increases due to losses within the system. Gain losses may also be part of the design specifications of the device.

More important than the gain decrease is a phase shift that occurs as frequency increases. When the phase shift reaches 90°, the amplifying system becomes critically unstable and oscillates. If, at 90° phase shift, the system gain is greater than one, each oscillation will cause output to increase in a manner illustrated in Figure 5.12b.

To eliminate this instability, the overall system gain is reduced so that at the 90° phase shift frequency, gain is less than unity. Oscillations are then damped out. Curve A in Figure 5.13 would be unstable at frequency f, whereas curve B, which represents an overall gain reduction, is stable. If the gain characteristic of the amplifier system is modified as shown in curve C, we obtain high gain and better performance (than B) at low frequencies while maintaining stability at high frequencies. This is achieved using a frequency sensitive network in the forward gain or feedback loops.

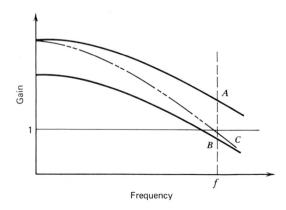

Figure 5.13   Amplifier gain characteristic modification.

5.6 Tooling Specification for NC

The command input for NC metal cutting machines specifies the coordinate location of some point on the cutting tool. Therefore, the accuracy with which the surface is generated depends upon the tool and tool holder dimensional tolerance. To illustrate this, consider a typical PTP drilling operation shown in Figure 5.14*a*.

The NC command sequence first moves the drill to a point above the workpiece surface. Then the drill is moved downward by an amount Δz, so that the desired hole depth is cut. If the NC command reference point is the tip of the tool, the depth increment Δz is the required hole depth plus the distance of the drill tip above the workpiece surface. A change in tool length will alter the distance between tool and workpiece, and if the NC command is not changed accordingly, the Δz movement will produce a hole of incorrect depth (Figure 5.14*b*,*c*).

Once the NC command sequence is in operation, the NC system has no way of determining the correct tool length. Therefore, errors in setup cause errors in surface generation. The gradual change in cutting tool dimensions due to tool wear during the metal removal process also affects the accuracy of the surface and requires adaptive control (Chapter Nine) for continuous compensation.

To eliminate problems that result from varying tool lengths, the MCU on certain NC machine tools contains a tool length compensation feature. When tool dimensions change, the operator can key into the MCU the required offset. The MCU automatically changes all NC program instructions to compensate for the dimensional variance.

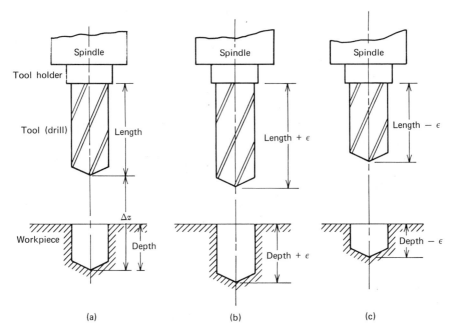

Figure 5.14 Tool length variance and dimensional accuracy.

Because the specification of NC instructions must *assume* tooling that is accurate to a given tolerance band, *qualified* or *preset* tooling is used with NC machines that require a high degree of accuracy.

Qualified tooling is more accurate than conventional machine tooling. Tool holders are manufactured to a guaranteed tolerance, usually ±0.02–0.04 mm. Cutting tools are also defined by prescribed tolerances. The NC part programmer* can use the qualified tolerances to generate NC instructions. When extremely tight tolerances are required, it may still be necessary to compensate for dimensional variance before the cut is made.

Preset tooling enables a high degree of precision to be maintained under industrial conditions by using special equipment to preset a tool in its holder. The tooling is designed so that adjustments can be made which will allow exact gaged dimensions to be attained.

It is important that tooling for NC machines be matched to the required part accuracy, since the output of the servomechanism is described by the tool. Errors in setup or tool dimension negate the most accurate machine control configuration.

---

* A discussion of NC part programming is presented in Chapter Seven.

5.7 Metrification and the NC System

The metric system of measurement (the S. I. system) will become the worldwide standard within twenty years. In 1976, only 10 percent of all manufacturers in the United States used the S. I. system. Legislation has provided the impetus for the initiation of the long awaited conversion from the English system.

The process of metrification requires more than converting English units to metric units. Standard component sizes, selected on the basis of whole or fractional English units, result in unwieldly decimal values in the metric system. For example, a simple dimensional specification of 1.5 in. ± 0.005 in., when converted literally to metric, becomes 38.1 mm ± 0.127 mm. A significant part of the conversion process concerns intelligent planning and design decisions to eliminate such unwieldly numeric values by a round-up or round-down process.

Numerical-control systems require adaption to the metric system. Many new NC machines already incorporate inch/millimeter features which enable the measurement system to be selected.

Three techniques are currently used to provide English/metric compatibility. It is important to recall the following relationships:

$$1 \text{ in.} = 25.4 \text{ mm}$$

$$1 \text{ mm} = 0.03937 \text{ in.}$$

$$0.0001 \text{ in.} = 0.00254 \text{ mm}$$

The first method of internal conversion makes use of a *dual line encoder* which has disks containing 500 and 635 lines per revolution with a resolution of 0.0001 in. Under English operation the system uses the 500-line disk, which multiplied by four produces 2000 pulses/revolution. Since each pulse corresponds to 0.0001 in., the system is capable of measuring a linear movement of 0.2000 in.

When operating in the metric mode, the encoder is switched to the 635-line disk. Once a factor of four has been applied, this arrangement produces 2540 pulses/revolution corresponding to a distance of 5.08 mm. In metric mode, electronic logic enables input to be changed from 0.0001 in. to 0.002 mm (the metric resolution). By switching to metric the resolution of the NC system is improved (0.002 mm = 0.0000787 in.).

The second method of metric conversion uses a *single line encoder* which feeds back data to a resolution of 0.002 mm. The error comparison circuitry contains logic that allows a numeric conversion to be performed on either the command or feedback signal so that a like-unit comparison can be made. For example, assume an input command of 0.0060 in. has been programmed. As the

NC system initiates motion, pulses corresponding to 0.002 mm begin to be counted. However, these pulses must be converted to represent 0.0000787 in. before comparison can be accomplished. After 76 pulses, a distance of 0.00598 in. has been traveled. Although a small error does occur, it is kept well within the system rated resolution.

The third method of conversion is used in NC machines that make use of a *softwired* (Chapter 10) MCU. Softwired NC replaces the conventional MCU with a programmable minicomputer. Such systems can provide for a broad range of inch-millimeter or other required conversions.

All of the methods described contain logic that provides for ipm and mpm feedrates. Eventually, all NC equipment will be manufactured only for S. I. usage. Until then, NC must be capable of supporting both systems of measurement.

References

1. Koenigsberger, F., *Design Principles of Metal Cutting Machine, Tools*, Pergamon, New York, 1964, pp. 42–68.
2. Sung, C. B., "Designing Machine Tools for Automatic Control," *The Engineer's Digest*, vol. 18, September 1957, pp. 391–98, and October 1957, pp. 436–38.
3. Dutcher, J. L., "Machine Tool Capacity," *Numerical Control in Manufacturing*, F. W. Wilson, ed., ASTME, 1963, pp. 217–29.
4. Boothroyd, G., *Fundamentals of Metal Machining and Machine Tools*, McGraw-Hill, New York, 1975.
5. Thomson, W. T., *Vibration Theory and Applications*, Prentice-Hall, Englewood Cliffs, N.J., 1965, pp. 228–42.
6. Welbourn, D. B., and Smith, J. D., *Machine Tool Dynamics—an Introduction*, Cambridge University Press, London, 1970.
7. Merritt, H. E., and Hohn, R. E., "Chatter—Another Control Problem," *Control Engineering*, vol. 14, December 1967, pp. 61–4.
8. Shinners, S. M., *Modern Control System Theory and Application*, Addison-Wesley, Reading, Mass. 1972, Chap. 3.
9. 't Mannetje, J. E., "Stabilizing Networks for Hydraulic Motors," *Control Engineering*, vol. 21, June 1974, pp. 55–8.

Problems

Note: Many problems make use of data contained in Table 5.2.

1. Explain why misalignment between a positioning slide and its ways may cause a point on the slide to move for a short distance in a direction opposite to the command direction. What dimensions affect the magnitude of this cocking motion?

Table 5.2
Physical Constants for Metals

Material	Modulus of elasticity, E, Pa $\times 10^{-10}$	Shear modulus, G, Pa $\times 10^{-10}$
Aluminum	7.1	2.52
Cast iron	10.0	4.14
Carbon steel	19.6	7.58
Stainless steel	19.0	7.30

2. Why is windup more severe in geared drive systems than in systems that make use of direct drive leadscrews?

3. A leadscrew of an NC positioning table has the following characteristics: diameter = 25 mm, pitch = 16 mm, length = 1 m; the material is carbon steel. If accuracy requires that lost motion be less than 0.004 mm, what is the allowable torque output of the servomotor?

4. The leadscrew described in Problem 3 is subjected to an axial force of 1000 N. Recalculate the allowable motor torque for the same degree of accuracy.

5. A positioning table is moved in the x and y directions by two hydraulically actuated pistons. Each piston is 0.6 m long, and made of carbon steel. The machine configuration is such that the x-piston diameter is 50 mm and the y-piston diameter is 0.25 mm. If both pistons experience a load of 5000 N along their respective axes, what will be the fixed coordinate location for the following PTP movement? Starting point—1.000, 1.000, x-increment command (Δx) = 75 mm, y-increment command (Δy) = 100 mm.

6. The machinable area for an NC machine (three-axis) is a volume $0.6 \times 0.6 \times 0.5$ m. All drive components are 40 mm in diameter and 1.0 m long, and made from stainless steel. Maximum torque in the horizontal and vertical directions is 15 N · m, and the positioning table has a mass of 100 kg. The maximum workpiece mass is 150 kg. Assuming that the horizontal and transverse axes are directly driven by leadscrews and that the vertical axis is powered by a hydraulic piston, what is the error due to windup and deflection along each axis?

7. Discuss the stick-slip phenomenon using plots of friction force vs actuator force, kinetic energy vs actuator force, and potential energy vs actuator force.

8. Using a good engineering design text as a reference, discuss the mechanics of the leadscrew and the relationships that are required to express axial force and input torque in terms of the geometric characteristics of the leadscrew.

9. Given the equations required for Problem 8, develop expressions that relate the windup and deflection of the leadscrew directly to its geometry.

10. An NC actuation system is modeled by the torsional pendulum shown in Figure 5.4b. If machine operating frequencies range from 10 to 60 Hz and the effective

inertia is 1000 kg · m · sec, what is the effective stiffness? If this stiffness were to be supplied by a circular steel shaft which is 16 mm in diameter, how long would the shaft be?

11. Using an engineering vibrations text as a reference, outline the Holzer method for obtaining the natural frequencies of a system modeled as shown in Figure 5.5. Discuss another method for the solution of this problem.

12. Using the Holzer method defined in Problem 11, show that the natural frequency for the geared system illustrated in Figure 5.6 is

$$\omega^2 = \frac{(n^2 J_M + J_L)K_1 K_2}{(K_2 + n^2 K_1)J_M J_L}$$

13. Use the Merchant metal cutting model to develop an expression for the velocity of the chip in terms of the shear and rake angles, ϕ and α, and the tool velocity, V.

14. Using the block diagram for chatter in Figure 5.8, determine the subsystem sensitivity with respect to cutting stiffness. Similarly, determine the subsystem sensitivity with respect to machine stiffness.

15. Discuss servo response to a step force (Figure 5.10) in terms of the system time constant, τ. How does the time constant relate to servo stiffness?

16. Given a double-acting hydraulic cylinder that drives a machine positioning table, plot frequency as a function of stroke length ($l = 0.2$ m to 0.9 m) for the following system specifications:

$$\text{oil bulk modulus} = 3.45 \times 10^8 \text{ Pa,}$$
$$\text{piston diameter} = 25 \text{ mm,}$$
$$\text{table mass} = 500 \text{ kg.}$$

Using a second ordinate plot cylinder stiffness vs stroke length, how are frequency and stroke length related?

17. Explain how an increase in loop gain will reduce valve drift. Justify your discussion with an example.

18. A servo valve is known to drift by 3 percent due to various internal parameters. To control this drift a velocity control loop with a gain of 200 is applied. What would be the velocity shift with the velocity control loop in operation?

19. Discuss some of the disadvantages of a servo loop that has extremely high gain. Relate your discussion to NC equipment.

20. The characteristic response of an amplifier is such that when a 90° phase shift has been reached, the gain equals 0.8. If the frequency at this point is 20 Hz and the signal amplitude is 1, sketch the response curve through 0.25 sec.

Chapter Six
NC System Input and Output

The numerical control process is designed around information exchange. For NC systems the origin and format of the information and the way in which it is processed must be defined in detail.

6.1 The Information Interface

During the numerical control process, symbolic data representing the motion of the NC machine as well as many operating commands are passed to the machine control unit (MCU). The content of this command information is in a standardized* symbolic format.

Symbols are represented by holes in punched tape or computer cards, magnetized domains on magnetic tape, or electronic impulses sent via computer communication lines. As long as the communication media is defined, an MCU can be designed to translate the information contained within the media and actuate servomechanisms to perform the required tasks.

6.2 Types of Media

The initiation of the NC process takes place by communication of symbols to the control unit. The ideal medium for communication and data storage packs information into a dense, easily interpreted code which after input to a reader is sent to the MCU at high speeds.

---

* A number of organizations propose standards, among them: National Bureau of Standards; Electronic Industries Association (EIA); International Standards Organization (ISO); Aerospace Industries Association (AIA).

There are four basic types of NC media: punched cards, punched tape, magnetic tape, and direct communication link with computer based information. Each type in this series represents an increased level of sophistication the use of which is dictated by the NC system under consideration.

6.2.1 Punched Cards

The punched card now has limited application as an input media for NC. However, since it was first originated by Herman Hollerith in 1887, the punched card has provided an efficient and accurate means of data storage and transfer. Editing of individual NC blocks (Section 6.4) is relatively simple, because a card may be extracted from the deck, retyped using a keypunch, and replaced. However, the punched card is a low density storage medium and can only be input to the MCU at rates that are much slower than other media types. The intrinsic ability to remove and replace cards also leads to the possibility of mis-sequencing or loss of an NC block (Reference 1).

The standard 80-column "IBM card" is 3.250 in. wide, 7.375 in. long, and 0.007 in thick. Hole size and spacing are precisely specified. Each card contains 12 rows of hole locations with 80 horizontal positions across the card.

Although any meaningful symbolic code could be devised for card input, the most common code is BCD (*Binary Coded Decimal*) or its extension, EBCDIC (*Extended Binary Coded Decimal Interchange Code*). The code for each character is represented by defining two sections on the card. The upper three rows are called *zone* rows, while the bottom nine represent the *digits* 1, 2, 3, . . . , 9. As illustrated in Figure 6.1, letter symbols are comprised of one

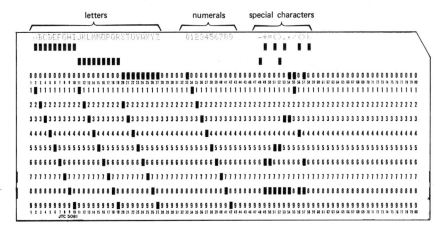

Figure 6.1   An 80-column *IBM* punched card.

zone and digit entry; numbers, by a digit entry alone; and special characters by 1, 2, or 3 punches as shown. The hole pattern for each character is unique.

Other types of punched cards are available, but they are less often used in numerical control applications. Remington Rand cards (45 and 90 columns) are the same size as the IBM cards, but have different size holes, spacing, and punch patterns. IBM has recently developed a smaller 96-column card for certain data processing applications.

6.2.2 Punched Tape

At this time in the evolution of numerical control systems, the majority of NC machines receives data from reels of 1 in. wide tape containing holes. The specifications for punched tape are standardized by the EIA and AIA, and are illustrated in Figure 6.2. Tapes are made of paper, paper-plastic sandwiches, aluminum-plastic laminates, or other materials, depending upon the characteristics of the tape reader and projected use of the tape.

Because tape rolls may be more than 1000 ft long considerable data can be stored on a single punched tape. It therefore functions as a more densely packed media than cards.

Paper tape perforators were first developed in the early 1940s, and have evolved to sophisticated units such as the *Flexowriter.** This automatic

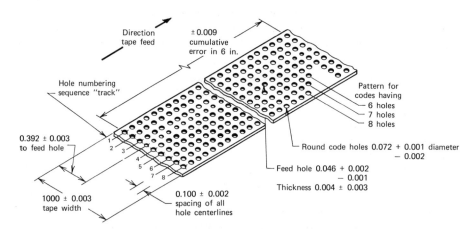

Figure 6.2 Specifications for perforated tape according to EIA standard RS-277 and NAS-943, NAS-955. (From Electronic Industries Association and Aerospace Industries Association. Reprinted with permission.)

*Tradename of Friden, Inc.

machine reads and prints tapes at high speed. Data is entered using a standard typewriter keyboard (with additional special function keys) and is converted into printed output and perforated tape. Figure 6.3 illustrates a typical configuration.

Punched tapes can also be generated directly from the output of computer based NC part-programming systems. A common approach makes use of magnetic tape as an output medium that is processed off-line by a minicomputer interfaced with an automated tape punch. This configuration, called a *media converter*, is used to transform data on magnetic tape to 1 in. punched tape.

Because data is stored on a continuous roll, only minor editing can be performed without the need to *splice* new data into the tape. During editing the hole codes must be interpreted visually, since no printed character representation exists on the tape. However, on-line computer editing techniques have been developed to enable rapid changes to be made. A new tape is automatically punched at the end of the editing session.

Input methods for punched tape require a tape reader whose input speed may vary from 10 to 1000 characters per second. The tape reading speed depends upon and is matched to the NC machine and the MCU. A typical NC system with tape preparation equipment, tape reader, MCU, and machine tool is illustrated in Figure 6.4.

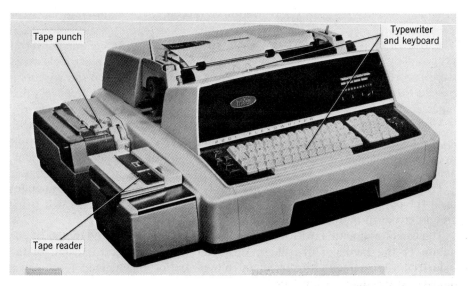

Figure 6.3 *Flexowriter* for preparation of punched tape (*courtesy of the Friden Division, Singer Co.*).

Figure 6.4 Components of a typical NC system (*courtesy Cincinnati Milacron*).

6.2.3 Magnetic Media

The use of a magnetic material as a recording and storage medium was proposed by Poulsen in the late 1800s. Sounds transduced into electrical energy were recorded on an iron wire by applying a magnetic flux to the wire as it passed through an electromagnet. It was not until the 1940s, however, that tapes coated with an iron oxide film were used to record magnetic pulses.

Modern magnetic tapes are made by coating a polyester or Mylar* film with an iron oxide coating. Such magnetic tapes have extremely high storage densities: 1600 to 6250 bytes per inch are common in computer applications, and higher densities are possible.

The symbolic code used to specify data is similar in many respects to codes used for paper tapes. However, magnetic tapes are always generated using computer based methods, and cannot be edited manually.

The use of magnetic tape for NC applications has been limited by the fragile nature of the media. Dirt, oil, or dust can cause *read* errors, and in the industrial environment ferrous metal filings and/or magnetized tools cause damage to the data on the tape.

* Tradename of E. I. duPont, Inc.

In recent years sealed magnetic tape cassettes, like those used in sound recording equipment, have been used with various NC systems. However, widespread application is not likely in the foreseeable future because a slight movement of the tape away from the reading heads causes a substantial reduction in signal and data input errors result.

A more reliable magnetic storage device, the flexible or *floppy* disk, has recently been introduced as an NC communication medium. A single disk contains the equivalent of about 2000 feet of punched tape in an area the size of a small phonograph record.

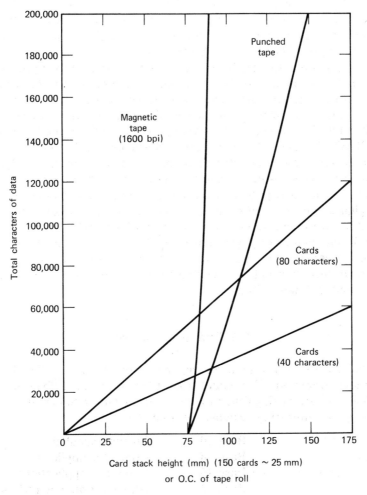

Figure 6.5 A comparison of media storage capacity.

Figure 6.5 graphically compares the storage capacity of input media relative to the height of a card deck and the outer diameter of a tape roll. The difference in storage capacities is immediately evident. A complete evaluation of physical NC input media also requires that characteristics of read rate, practicality, and cost be considered.

6.2.4 Computer Transmitted NC Data

Although punched tape is used in the majority of NC applications, methods linking a digital computer directly to the NC machine have been developed. Computerized Numerical Control (CNC) and Direct Numerical Control (DNC) are outgrowths of this technology. Both systems are part of the computer-aided manufacturing (CAM) concept (Chapter Ten).

In a CNC system, the machine control unit is replaced by a programmable minicomputer. A number of NC programs may be stored in the computer's memory. After the NC instructions are stored, however, the minicomputer performs most or all of the basic numerical control functions. Because data is internally stored and accessed, the use of physical media is greatly reduced.

The DNC environment is the most sophisticated of NC systems. A large scale computer is used to control many NC machines simultaneously. All NC programs are stored using peripheral devices which have very large storage capacities. A computer program executed by the central processing unit (CPU) of a digital computer is often used to replace the MCU, and can send NC information to the machines at rates as high as 160,000 bits/sec.

With the elimination of physical media and the introduction of the computer as a control device, many additional monitoring capabilities are possible. An extension of this philosophy is a computer integrated manufacturing system to be discussed in Chapter Ten.

6.3 Symbolic Codes

A symbolic code is a system of characters, numbers, or symbols used to represent information. The codes that have been developed for storing numerical data are distinguished by the *radix* or *base number* of the system. Four common numbering systems are used:

Decimal	Base 10	Manual calculations
Octal	Base 8	Computer applications
Hexidecimal	Base 16	Computer applications
Binary	Base 2	All electronic control/computing equipment

Although all of these number systems are important in computer applications, the binary number system is used for NC data codes. The control unit of an NC machine contains electronic circuitry that responds to either of two conditions, *on* or *off*. This *true-false* logic is ideally suited to a number system that has only two admissible *marks*. The marks in a binary system are *0* and *1*; hence, two states can be easily and efficiently represented.

A binary digit (abbreviated *bit*) is a mark of a number system employing two distinct kinds of characters. Table 6.1 illustrates the numbers that can be represented using four bits.

Table 6.1
Binary Numbers

Decimal	Binary	Power of 2
0	0000	
1	0001	2^0
2	0010	2^1
3	0011	
4	0100	2^2
5	0101	
6	0110	
7	0111	
8	1000	2^3
9	1001	
10	1010	
11	1011	
12	1100	
13	1101	
14	1110	
15	1111	
16	(1) 0000	2^4

By combining a string of marks, any number can be represented. For example, the binary number

$$101101 = 2^5 + 2^3 + 2^2 + 2^0$$
$$= 32 + 8 + 4 + 1$$
$$= 45 \text{ (decimal)}$$

A pure binary number system is rarely used for NC input. Instead a more sophisticated BCD or ASCII symbolic code is most frequently used.

6.3.1 An Example of Binary Expansion

As an example of binary number representation, we can write the decimal number 1754 as a binary string.

From Table 6.2, the appropriate powers of 2 can be obtained. Hence, from the table

$$1754 = 1024 + 512 + 128 + 64 + 16 + 8 + 2$$

or in binary code: 11011011010.

If the decimal and binary numbers are divided by 2, the numbers become

$$877 = 1101101101$$

Examining each binary string, it should be noted that dividing by a power of two is equivalent to dropping the last mark, or *shifting right* one bit.

```
1  1  0  1  1  0  1  1  0  1  0
 ↘  ↘  ↘  ↘  ↘  ↘  ↘  ↘  ↘  ↘        Division by 2
   1  1  0  1  1  0  1  1  0  1
```

Table 6.2
Powers of Two

Power of 2	Mark position	Quantity
0	1	1
1	2	2
2	3	4
3	4	8
4	5	16
5	6	32
6	7	64
7	8	128
8	9	256
9	10	512
10	11	1024

6.3.2 Binary Coded Decimal

Eight track punched tape is the most common input media for NC systems. Hence, all data in the form of symbols, letters, and numbers must be representable by eight marks. The BCD code has been devised to satisfy this requirement.

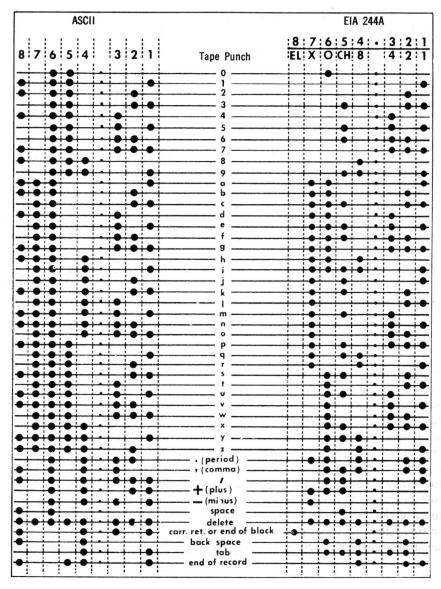

Figure 6.6 EIA and ASCII codes for perforated tape used in NC applications (*with permission of Electronic Industries Association. From NC/CAM Guidebook and Directory*, copyright © 1976 by Gardiner Publications. Reproduced with permission).

Using BCD coding, the numerals 0 through 9 are specified using only the first four tape tracks. Letters, symbols, and special instructions are indicated by using tracks 5 through 8 in combination with the numeral tracks. A complete BCD character set, based on EIA Standard RS-244A, is illustrated in Figure 6.6.

For punched tape each bit, or mark, is represented by the presence or absence of a hole in the tape. Although tape punching equipment is reliable, there is always a possibility that a hole will be erroneously added or deleted. To aid in the detection of such errors, a *parity bit* is added to the code.

Each BCD character must have an odd number of holes. By punching a parity bit along with all even bit strings, all characters have an odd number of holes. If an even number of holes is detected, it is by definition an error, and a *parity check* occurs. This simple method provides some protection from input errors. For BCD code, track 5 is used to specify the parity bit. In Figure 6.6, the bit is *on* for numbers 3, 6, 7, 9 because each binary representation by itself contains an even number of bits.

The remaining BCD characters are represented as shown in the figure. For example a single hole in track 6 represents a zero, and track 8 is used to indicate *end-of-block (eob)*.

6.3.3 ASCII

ASCII (*American Standard Code for Information Interchange*) was formulated to standardize punched tape codes regardless of application. Hence, ASCII is used in computer and telecommunications as well as in NC applications. ASCII code was devised to support a large character set that includes upper and lower case letters and additional special symbols not used in NC applications.

Figure 6.6 illustrates the ASCII subset applicable to NC. The differences between it and EIA RS-244A are many. A tape reader which can read EIA code cannot read ASCII without extensive modification. Hence, there is some resistance towards using ASCII. A new EIA standard, RS-358, incorporates a subset of the ASCII codes used for numerical control work, and many new control systems now accept both codes. It is likely that the move toward ASCII standardization will progress as older NC equipment is replaced.

6.4 Tape Input Formats

Numerical control information is passed to the MCU in block format. Each block of NC data may be arranged differently, depending on the control system requirements of the system configuration. Four basic tape formats are used for

NC input:

1. *Fixed sequential* format.
2. *Block address* format.
3. *Tab sequential* format.
4. *Word address* format.

Regardless of format, each NC block must be capable of specifying *dimension* and *nondimension* data. Dimension data comprises linear and angular motion commands for the machine. Nondimension data are the preparatory functions used to describe specific types of movement; the miscellaneous functions which control machine operation; sequence information; feeds and speeds; and tooling specifications. By convention, data within an NC block is specified in the following order:

$$\mathbf{n\ g\ xyzab\ f\ s\ t\ m}\ eob \qquad (6.1)$$

where

$$\mathbf{n} = \text{sequence number}$$

$$\mathbf{g} = \text{preparatory function}$$

$$\mathbf{xyzab} = \text{dimension data}$$

$$\mathbf{f} = \text{feed function}$$

$$\mathbf{s} = \text{speed function}$$

$$\mathbf{t} = \text{tool function}$$

$$\mathbf{m} = \text{miscellaneous function}$$

$$eob = \text{end-of-block}$$

Fixed sequential tape format requires that each NC block be the same length and contain the same number of characters. This restriction enables the block to be divided into substrings corresponding to each of the data types specified above. Since block length is invariant, all values must appear. For example, even if feed and speed are the same for 100 blocks of NC data, **f** and **s** must be coded in each block. All values are defined using a predefined number of digits.

Block address tape format eliminates the need for specifying redundant information in subsequent NC blocks through the specification of a *change code*. The change code follows the block sequence number and indicates which values are to be changed relative to the preceding blocks. All data must contain a predefined number of digits.

Tab sequential tape format uses a special symbol called the *tab* to separate data values within a block. Two or more tabs immediately following one another indicate that the data which would normally occupy the *null* locations is redundant and has been omitted.

Word address tape format is the only input scheme that uses alphanumeric (letters and numbers) data specification. Each data value is preceded by a letter, which indicates the type of data that follows. Hence, redundant information is merely omitted along with the appropriate letter. In some types of controls using word address format, leading or trailing zeros may be omitted, thereby shortening the block length.

Tab sequential and word address are the most prevalent tape formats. The following NC block representations illustrate the use of these two formats.

Tab sequential:

t001　t01　t07500　t06250　t10000　t612　t718　t　t　eob

t002　t　t08725　t06750　t　t　t　t　eob

t003　t　t　t　t05000　t520　t620　t01　t　eob

where t represents a tab.

Word address:

n001　**g**01　**x**07500　**y**06250　**z**10000　**f**612　**s**718　*eob*

n002　**x**08752　**y**06750　*eob*

n003　**z**05000　**f**520　**s**620　**m**01　*eob*

Both formats specify exactly the same machine tool operation using data order described in expression (6.1). Since fewer characters are required to define the NC operation using a word address format, shorter tapes requiring less programming result.

6.5 Communications with the MCU

The MCU is a special purpose computer which executes commands stored in its memory. The machine control has a storage register for each piece of information contained in an NC block. As the tape is read, data is placed in the appropriate register using the methods schematically illustrated in Figures 6.7 and 6.8.

Each tab (tab sequential tape format) actuates an electronic switch which passes data into the appropriate storage register. If multiple tabs are encountered, the switch bypasses the designated number of register paths, leaving the data in the register unchanged. In this manner, redundant data is saved from one block to the next. When an *eob* character is encountered the information block is ready for execution and the register switch is reset to the first path. Figure 6.7 illustrates this process (Reference 2).

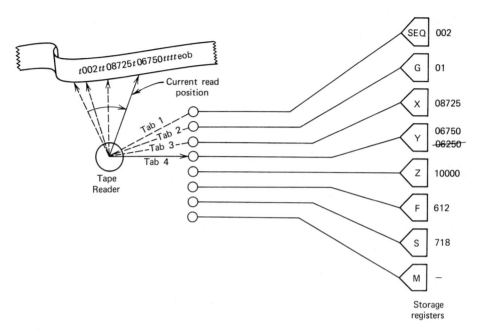

Figure 6.7 Schematic representation of tab sequential input to the MCU.

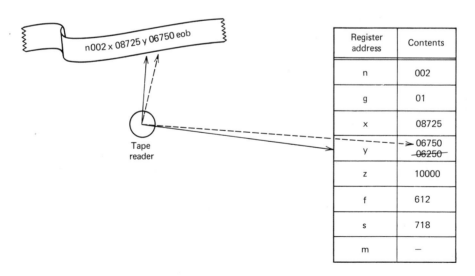

Figure 6.8 Schematic representation of word address input to the MCU.

165

Control units which accept word address formatted blocks contain circuitry that interprets the character preceding numeric data and routes the data to the appropriate storage register. For this reason, word address data may be specified in any order* since registers are filled based on the address only. A schematic representation is illustrated in Figure 6.8.

6.6 Basic NC Input Data

The command data contained in an NC block is obtained using methods that range from referring to simple tabular information to performing rather complex mathematical calculations. Calculations necessary for dimension data are presented in Chapter Seven.

Nondimension values take on two forms: (1) functions that are selected based on predefined tabular standards and (2) functions that must be computed based on the operation to be performed. The first group consists of four functions: sequence number, and preparatory, miscellaneous, and tool functions.

6.6.1 Sequence Numbers

The sequence number (**n** code) is used to identify each block within an NC program and provides a means by which NC commands may be rapidly located. Some control units require that sequence numbers be input in ascending order, whereas other systems allow any three digit number to appear after the **n** symbol. Hence, the NC programmer can develop a code which indicates the specific function for a given block.

6.6.2 Preparatory Functions

The preparatory function (**g** code) is used as a communication device to prepare the MCU. The **g** word indicates that a given control function, such as linear interpolation (**g**01), is to be requested. The number codes for preparatory functions have been standardized in EIA Standard RS-273. A table of preparatory function codes is listed in Table 6.3.

6.6.3 Miscellaneous Functions

Miscellaneous functions (**m** code) are used to designate a particular mode of operation for a numerically controlled machine tool. Most miscellaneous

---

* In standard practice, the order corresponds to expression (6.1).

Table 6.3
Preparatory Functions (EIA Standard RS-273)

Code	Function
g00	Point to point, positioning
g01	Linear interpolation
g02	Circular interpolation arc CW
g03	Circular interpolation arc CCW
g04	Dwell
g05	Hold
g06 & g07	Unassigned
g08	Acceleration
g09	Deceleration
g10	Linear interpolation (long dimensions)
g11	Linear interpolation (short dimensions)
g12	Unassigned
g13–g16	Axis selection
g17	xy plane selection
g18	zx plane selection
g19	yz plane selection
g20	Circular interpolation arc CW (long dimensions)
g21	Circular interpolation arc CW (short dimensions)
g22–g24	Unassigned
g25–g29	Permanently unassigned
g30	Circular interpolation arc CCW (long dimensions)
g31	Circular interpolation arc CCW (short dimensions)
g32	Unassigned
g33	Thread cutting, constant lead
g34	Thread cutting increasing lead
g35	Thread cutting, decreasing lead
g36–g39	Reserved for control use only
g40	Cutter compensation cancel
g41	Cutter compensation—left
g42	Cutter compensation—right
g43–g49	Cutter compensation if used; otherwise unassigned
g50–g59	Unassigned
g60–g79	Reserved for positioning only
g80	Fixed cycle cancel
g81	Fixed cycle 1
g82	Fixed cycle 2
g83	Fixed cycle 3
g84	Fixed cycle 4
g85	Fixed cycle 5
g86	Fixed cycle 6
g87	Fixed cycle 7
g88	Fixed cycle 8
g89	Fixed cycle 9
g90–g99	Unassigned

functions deal with opposing machine conditions; for example coolant *on*, coolant *off*. The number codes for miscellaneous functions are listed in Table 6.4.

Table 6.4
Miscellaneous Functions (EIA Standard RS-273)

Code	Function
m00	Program stop
m01	Optional (planned) stop
m02	End of program
m03	Spindle CW
m04	Spindle CCW
m05	Spindle OFF
m06	Tool change
m07	Coolant No. 2 ON
m08	Coolant No. 1 ON
m09	Coolant OFF
M10	Clamp
m11	Unclamp
m12	Unassigned
m13	Spindle CW & coolant ON
m14	Spindle CCW & coolant ON
m15	Motion +
m16	Motion −
m17–m24	Unassigned
m25–m29	Permanently unassigned
m30	End of tape
m31	Interlock bypass
m32–m35	Constant cutting speed
m36–m39	Unassigned
m40–m45	Gear changes if used; otherwise unassigned
m46–m49	Reserved for control use only
m50–m99	Unassigned

6.6.4 Tool Function

The tool function (**t** code) is used in conjunction with the miscellaneous function for tool changes (**m06**), and as a means of addressing the new tool. Numerical-control machining centers often have an automatic tool change apparatus for supplying the tool designated using the **t** word. For example, if

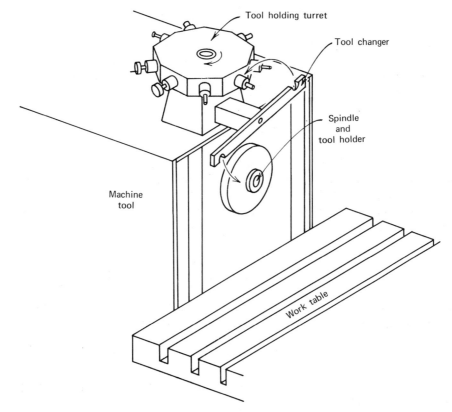

Figure 6.9 Turret arrangement for automatic tool change.

the tool change device consists of a turret arrangement, illustrated schematically in Figure 6.9, the NC blocks **m06** *eob* and **t04** *eob* indicate (1) that a tool change is desired and (2) that position 4 on the turret should supply the tool. The tool function designation can consist of numbers of up to five digits in length.

The remaining nondimension words for feed and speed functions require calculations based on control system requirements and operational limitations. Two coding systems are generally available:* the EIA *Magic-3* Code, and the EIA *Inverse Time* Code. Coding systems are used so that any feed or speed value, regardless of magnitude, can be denoted using only three digits.

---

* Other systems based on inverse time, are used by a number of MCU manufacturers.

6.6.5 The Magic-3 Code

To encode data using the *Magic-3* code, the decimal point of the desired feedrate or speed is moved to precede the first nonzero digit. The number is then rounded to two significant digits following the decimal point and expressed in scientific notation. For example, consider a feedrate of 2.44 mpm. Following the encoding procedure

$$2.44 = .24 \times 10^1$$

The first digit of the *Magic-3* code represents the power of 10 resulting from the above expression *plus* 3. Hence, the name *Magic-3*. For a feedrate of 2.44 mpm, the first digit of the code would be $1 + 3 = 4$. The remaining two digits are the rounded down digits. Therefore, 2.44 mpm becomes 424 in *Magic-3* code. The *Magic-3* code is only valid within finite limits.

6.6.6 Examples to Illustrate the Magic-3 Code

Using the *Magic-3* code, the feedrates 1.645 mpm, 0.725 mpm, and 0.050 mpm can be represented by three digit numbers. Considering the first value

$$1.645 \text{ becomes } .16 \times 10^1$$

Hence, 1 and 6 are the second and third *Magic-3* digits, and 4 (exponent 1, plus 3) is the first digit. Therefore,

$$1.645 \text{ mpm} \rightarrow 416 \ (Magic\text{-}3)$$

Likewise, 0.725 mpm becomes 372, and 0.050 mpm becomes 250 in *Magic-3* code.

The values that can be represented using *Magic-3* are limited by the code definition. The first *Magic-3* digit must be zero or positive, and since the second digit may not be zero unless the number is zero, we have the smallest code number 010. Hence, the exponent must be $0 - 3 = -3$, and the smallest number is

$$.10 \times 10^{-3} = .00010$$

The largest first digit is nine. Hence, 999 would be the largest code value. Using the same method as above the largest number is

$$.99 \times 10^6 = 990000$$

6.6.7 Inverse Time Code

Inverse time coding is used exclusively for feedrate, f, codes. It provides a feedcode which is equal to the reciprocal of the interpolation time in minutes. Depending on the MCU, linear and/or circular interpolation are available for continuous path NC machines. Hence, the inverse time feedcode, F_c, can be expressed as

$$F_c = \frac{1}{t} = \frac{V}{s} \tag{6.2}$$

or

$$F_c = \frac{\theta}{t} = \frac{V}{R} \tag{6.3}$$

where equation (6.2) applies to linear interpolation and equation (6.3) to circular interpolation, and V is the axis velocity for linear interpolation or the feedrate for circular interpolation, s is the distance between points, R is the arc radius, θ is the arc length (radians), and t is the time.

Consider a machining operation that requires the tool to move in a straight path for 0.3 m, at a feedrate of 1.1 mpm. From equation (6.2)

$$F_c = V/s$$
$$= 1.1/0.3 = 3.6/\text{min}$$

The EIA Standard RS-274A specifies that the feed command for inverse code may range from 0.1 to 999.9 ipm (English units). Therefore, four digits are required for specification of the f code when using inverse time. From the above example, the feedcode would be f0036.

Inverse time code has an important disadvantage. For low feedrates, the percent error in specification can be large. Consider a tool traveling on a 0.4 m path at 0.1 mpm. From equation (6.2) the feedcode would be

$$F_c = \frac{0.1 \text{ mpm}}{0.4 \text{ m}} = .250/\text{min}$$

Using standard feedra ᷈ coding, the feedrate word is

f0002 or f0003

Regardless of the rounding choice, a 20 percent error occurs. This problem is resolved by using increased resolution inverse time coding. Although more characters may be required for the feedword,* the errors for low feedrates are eliminated (Reference 3).

* Many controls do not require leading zeros. Hence, f0002 could be represented as f2.

The maximum and minimum time spans for linear interpolation can be computed based on standard inverse time code. By definition of inverse time

$$F_c = \frac{1}{t}$$

or

$$t = F_c^{-1}$$

From the specifications in the text

$$\frac{1}{999.9} \leq t \leq \frac{1}{0.1}$$

Therefore, $0.001 \, \text{min} \leq t \leq 10 \, \text{min}$ is the time range for linear interpolation between points.

6.7 Function Specification—Some Considerations

6.7.1 Feedrate

When a feedrate is chosen, the part programmer must consider the control system characteristics so that the tool path that is input to the MCU is the same path traced by the NC machine. Aside from the problems of feedrate specification based on workpiece machinability, the most significant feature in feedrate choice is the need to adjust the acceleration/deceleration characteristics to prevent overcut or undercut.

Some modern NC contouring machines use a servo system that operates with a *time lag* between the command coordinate and the actual tool position. In these *soft* or low gain servo systems, the tool may lag the commanded position by as much as 0.0125 to 0.0750 mpm/0.02 mm, with the result that stepping to programmed feedrate at a corner causes the undercut illustrated in Figure 6.10. Two solutions can be applied. A *dwell block* may be programmed, allowing the machine tool to approach the corner at low velocity. This reduces undercut to acceptable amounts. Also, the tool path can be *looped* in a manner that eliminates undercut (Figure 6.11). Looping requires additional cutter commands and may be restricted by workpiece or tool geometry.

For given servo characteristics, undercut is a function of feedrate and cornering angle. Studies (Reference 4) have shown that cornering error can be one to two orders of magnitude greater than the resolution of the machine tool. Test data indicates that a machine tool with 0.002-mm resolution taking a 90° corner at 1.5 mpm generates an error of 0.5 mm, which is 250 times the resolution of the machine.

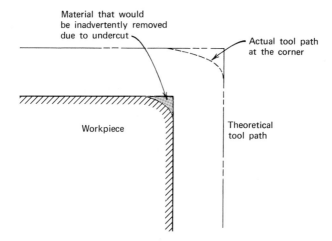

Figure 6.10 Cornering undercut due to control system lag.

The calculations necessary to determine the actual cutter path are impractical to perform manually for each corner. During computer-assisted NC programming these calculations are performed using a computer program called the *postprocessor*.

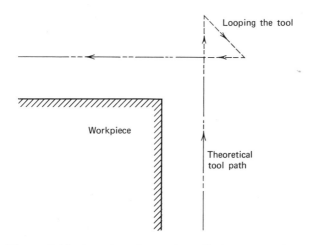

Figure 6.11 *Looping* the tool to eliminate undercut.

6.7.2 Undercut Error

When NC blocks are generated manually, a rough approximation of undercut error can be developed. Considering a corner angle θ, and a feedrate, f, the tolerance value

$$t' \sim |\cos \theta - f/(n \cdot l)| \tag{6.4}$$

where t' = tolerance (approximate); l = lag unit, generally ranging from 0.0125 to 0.0750 mpm/0.02 mm; and n = 1000 or 10000 depending on $\cos \theta$.

If $t' \ll t$ (the actual tolerance), no corrective action is necessary. If however $t' \approx t$, a dwell block should be programmed. Equation (6.4) was empirically derived as an approximate indicator of undercut (Reference 3).

When the machine is rapidly traversing toward a point to commence cutting, a deceleration block must be introduced to insure that the tool does not overshoot. Overcut on inside corners can also be eliminated using proper deceleration.

The overcut error, e, can be approximated using the following relation:

$$e = \frac{\Delta v}{120 \pi B} \tag{6.5}$$

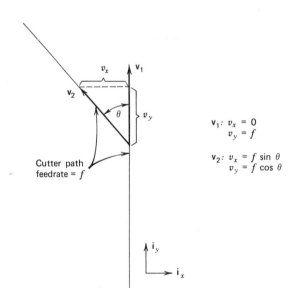

Figure 6.12   Velocity vectors at cornering point on tool path. Feedrate along path is constant.

where B is the servo bandwidth, and Δv is the velocity step of the incoming and outgoing velocity vectors. From Figure 6.12, for f, the path feedrate,

$$\mathbf{v}_1 = 0\mathbf{i}_x + f\mathbf{i}_y$$
$$\mathbf{v}_2 = f \sin \theta \mathbf{i}_x + f \cos \theta \mathbf{i}_y$$

Therefore, for overshoot in the y direction

$$\Delta v = v_{1y} - v_{2y} = f(1 - \cos \theta)$$

Expression 6.5 indicates the approximate overshoot error for a given motion.

6.7.3 An Example to Compute Overshoot

A cutting tool is required to move along the line segments defined by points $(0, 0)$, $(10, 0)$, and $(15, 3)$. If the incoming feedrate is programmed at 0.5 mpm and the outgoing feedrate is 0.65 mpm, approximate overshoot at the corner can be determined. The machine tool servo bandwidth is 10 Hz.

To determine the overshoot, equation (6.5) is used. The turning angle for the tool path

$$\theta = \tan^{-1}\left(\frac{0-3}{10-15}\right) = \tan^{-1}(.6)$$

$$\theta = 30.9°$$

and

$$\mathbf{v}_1 = f_1\mathbf{i}_x + 0\mathbf{i}_y = 0.5\mathbf{i}_x$$

$$\mathbf{v}_2 = f_2 \sin \theta \mathbf{i}_x + f_2 \cos \theta \mathbf{i}_y$$
$$= 0.65 \sin (30.9)\mathbf{i}_x + 0.65 \cos (30.9)\mathbf{i}_y$$
$$= 0.333\mathbf{i}_x + 0.557\mathbf{i}_y$$

For overshoot in x,

$$\Delta v_x = v_{x1} - v_{x2} = 0.5 - 0.333 = 0.167$$

Then, from equation (6.5)

$$e = \frac{\Delta v}{120\pi B} = \frac{0.167}{120\pi(10)} = 0.044 \text{ mm}$$

6.7.4 Constant Surface Speed

Once the feedcode has been established in a milling operation, the tool moves across the workpiece at the required *surface speed* for all linear movements. This situation becomes more complicated for NC turning operations. The lathe

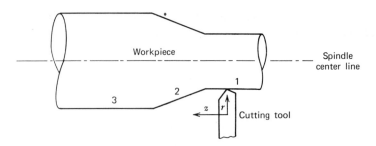

Figure 6.13 Surface speed for different turned surfaces.

tool will achieve a constant surface speed only when the tool position maintains a constant radius from the workpiece centerline.

The surface speed v_s for a turning operation is defined as

$$v_s = r\omega \tag{6.6}$$

where r is the radius at a given instant of time, and ω is the angular velocity, $\Delta\theta/\Delta t$. Referring to Figure 6.13, v_s remains constant as the tool moves along surface 1 in the z direction. As surface 2 is engaged each Δz results in an increase in radius, Δr, and from equation (6.6)

$$\Delta v_s = \Delta r\omega$$

or the new surface speed

$$v'_s = v_s + \Delta v_s = (r + \Delta r)\omega \tag{6.7}$$

Therefore, to maintain a constant surface speed (CSS), the angular velocity must be changed. This is done by dividing the tool feedrate in the r direction into small increments and changing ω accordingly. In this manner the rpm to mpm ratio for constant surface speed can be maintained.

The incremental method accuracy depends upon the number of steps taken, and results in longer program tapes. The incremental speed method can be eliminated if the NC lathe is equipped with a CSS control loop that takes the desired surface feedrate, f, as input via a function word in an NC block. The spindle of the lathe is equipped with an incremental encoder, or oscillator, that generates a frequency equal to $\Delta\theta/\Delta t$, the angular velocity. This multiplied by the radial dimension indicates the actual surface speed, v_s, at any instant of time. The required surface speed is expressed as

$$v_{\text{required}} = \Omega r f$$

where Ω is an oscillator fixed at a precalculated angular velocity frequency, r is the radial dimension (m), f is the programmed feedrate (mpm).

The CSS control loop compares the v_s and $v_{required}$ frequencies. If the two are equal, the present spindle speed is maintained. If $v_s \neq v_{required}$, corrective action is taken (Reference 5).

6.8 Verification of NC Input

Once the NC part program, consisting of blocks of dimension data and nondimension functions, has been generated, a technique called tape* verification is used to insure that NC commands are correct. Although machining and inspecting an actual part is the only complete guarantee that the data contained in an NC part program is error free, two basic types of tape verification methods have been developed. These are (1) machining substitute material and (2) graphic proofing.

6.8.1 Machining Substitute Materials

Numerical control is often used to manufacture parts made from expensive workpiece materials that exhibit poor machinability. Cost, availability, and machining time frequently prohibit the use of the actual part material for the test piece. For this reason, an NC part program is verified using a *substitute material*.

Figure 6.14   Substitute material for NC tape proofing (*courtesy Goldenwest Products*).

* We use tape as a synonym for any NC communication media.

Light metals, plastics, foams, laminates, wood, and other castable low cost materials are used for NC tape proofing. In some cases, the actual part material, salvaged from the scrap pile, is used. Figure 6.14 illustrates a complex component NC machined from a closed cell, foam plastic substitute material. The benefits of using substitute materials include lower cost, improved machinability, and shorter machining times (Reference 6).

6.8.2 Graphic Proofing

Errors in NC commands have many origins. These include drawing misinterpretations, mathematical errors, typing (and format) mistakes, and communication equipment malfunctions. However, these errors can be detected *before* machine tool tests are begun by *graphic proofing* techniques.

Graphic proofing provides a visual representation of the cutting tool path. This representation may be a simple two-dimensional plot of cutter center line path or a dynamic display of tool motion using computer generated animation.

There are three major categories for graphic proofing equipment:

1. Automatic drafting machines.
2. NC tape verifier/editors.
3. Cathode ray tube (CRT) devices.

The most widely used method of proofing NC tapes is the automatic drafting machine (ADM). The ADM is a numerically controlled device which plots the cutter center line path in two simultaneous axes of motion. The complexity, size, and scope of ADM systems vary widely. Some systems are combined with computer programs which produce orthographic plots of three dimensional motion and/or perspective plots.

The size of such systems range from a 0.280 m wide drum plotter to *flat-bed* devices which measure 2 m long by 6 m wide. Scaling of plots enables the representation of any cutter path.

NC tape verifier/editors combine a paper tape reader with a noninteractive CRT display unit which has limited editing facilities. Such devices read punched paper tape directly and can be used to diagnose gross NC data errors. The verifier provides the part programmer with a compact device suitable for rapid detection of major errors. Its small screen size and limited hard copy capabilities make close tolerance verification difficult.

The highest level of sophistication is the cathode ray tube (CRT) display. Some CRT systems display the NC data in a similar manner to the ADM. However, the real benefit of the CRT is realized when the displays are *interactive*.

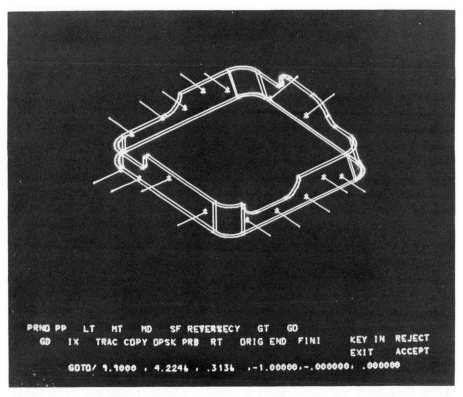

Figure 6.15 NC tape generation and verification using interactive computer graphics (*courtesy McDonnell Douglas Corporation*).

An interactive graphics system is linked directly to a host computer and contains software which enables the part programmer to communicate with the display. The possibilities of interactive graphics systems go far beyond tape proofing. Systems currently in full production use, such as shown in Figure 6.15, provide capabilities that allow the part program to be developed with the aid of the computer, debugged by *running* the cutter on the screen, and finally edited if necessary.

References

1. Tholstrup, H. L., "Information Processing and Storage," *Numerical control in Manufacturing*, F. W. Wilson, ed., ASTME, 1963, pp. 371–95.
2. Roberts, A. D., and Prentice, R. C., *Programming for Numerical Machines*, McGraw-Hill, New York, 1968, pp. 68–84.

3. _____. *N/C Handbook*, 3rd ed., L. J. Thomas, ed., Bendix Corporation, Industrial Controls Division, 1971, pp. 113–18.
4. Burkey, R. M., and Broadwell, W. B., "Dynamic Model for Contouring NC Devices," *Proceedings of the Ninth Annual Technical Conference*, Numerical Control Society, 1972.
5. _____. "Constant Surface Speed," *N/C Application Guide*, Allen-Bradley Company, Systems Division.
6. Utterback, L., "Tape and Tool proofing for Numerical Control," *AIA Report*, Project MC121, October 1973.

Additional References

Childs, J. J., *Principles of Numerical Control*, The Industrial Press, New York, 1965.
Dyke, R. M., *Numerical Control*, Prentice-Hall, New York, 1967.
Olesten, N. O., *Numerical Control*, Wiley-Interscience, New York, 1970.

Problems

1. Given the dimensions for punched cards and tape discussed in the text, calculate the character density. Discuss the relative merits of cards and tape.
2. The parity bit is a simple method for tape error checking. Why isn't each character checked for validity based on all holes punched? Would this be possible?
3. Could a five-track punched tape be used to represent any numeric string, and would all letters and numbers be represented by this tape? Explain your answer.
4. A given MCU uses word address format for input. It requires the following addresses: **x, y, n, f, s**, and *eob*. Could a five-track punched tape be used with such an MCU? If so, devise your own code to represent the word addresses.
5. Consider a DNC system as described in this chapter. If a computer program replaces the MCU, schematically illustrate (use a block diagram) the inputs to the MCU program and the outputs which it generates. Don't forget machine function monitoring.
6. The register in an MCU contains the binary number 101101001001. What is the decimal equivalent?
7. Given the binary number 1011010000111010110, what binary number results if we divide by 2? By 4? What binary number results if we multiply by 2?
8. A certain coding system requires that 128 characters be represented. What is the smallest bit string that could be used for such a system?
9. Using ASCII code, show what a punched tape would look like for the following blocks of word address data.

$$\text{n001 x12500 y03750 f241 s322 } eob$$
$$\text{n002 y07500 } eob$$

10. Decode the EIA-244A code given in Figure P6.1. Tape format is tab sequential.

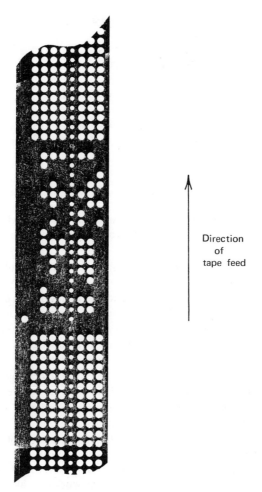

Direction
of
tape feed

Figure P.6.1   Problem 6.10.

Information format is (**n, x, y, f, s,** *eob*). Use EIA code to regenerate the data using word address format. Which section of tape is longer?

11. Figures 6.7 and 6.8 illustrate the methods used to decode tab sequential and word address format tapes. Illustrate a method for decoding fixed sequential input. Hint: A character counter will be necessary.

12. Given the NC part program below, draw the tool path indicating all feedrates, speeds (decimal values), miscellaneous and preparatory functions invoked before or on a given path. Tape format is word address. Dimension data is given to three decimal places.

n001 g01 x1000 y2500 *eob*
n002 m13 *eob*
n003 y2750 f416 s320 *eob*
n004 x2000 y3500 *eob*
n005 g05 *eob*
n006 g01 x2500 y3000 *eob*
n007 g05 *eob*
n008 g01 y2500 f425 *eob*
n009 x1000 *eob*
n010 m09 *eob*
n011 m05 *eob*
n012 m06 t04 *eob*
n013 m02 *eob*

13. What is the percent error between actual and coded value of a 0.15 mpm feedrate if *Magic-3* coding is used? Compare this value to EIA inverse time coding for the same feedrate.

14. Explain the benefits derived from using an increased resolution inverse time feedcode. What possible drawbacks does such a system have?

15. Discuss the parameters that must be considered to develop an exact expression for a cutting tool path as it moves through an angle θ. Develop the necessary equations.

16. For the tool path described by blocks 3 and 4 of the NC part program of Problem 12, is a dwell block necessary to eliminate overcut if the required tolerance is 0.01 mm? Show all calculations to justify your answer.

17. If the machining tolerance is 0.01 mm, is the dwell block (n005) of the NC part program of Problem 12 really necessary? Justify your answer.

18. Derive an expression for spindle speed, ω, as a function of tool feedrate in the radial direction so that surface speed remains constant.

19. The spindle of an NC lathe is turning at 400 rpm. A cutting tool moves along a path machining a cone whose cross section makes an angle, α, with the workpiece center line. The tool begins at a radius of 0.1 m. If the tool feed parallel to the center line is 0.1 mpm and the tool feed into the work (i.e., along the radial axis) is 0.0125 mpm, what is the starting surface speed? What is the surface speed after 3 min?

20. Given the machining conditions of Problem 19, what intervals should be taken to maintain a surface speed within 10 percent?

21. In certain machining situations, tool deflection affects machining accuracy. Can a substitute material be used to evaluate tool deflection? Explain your answer from an engineering viewpoint.

The following problems are suggested for those students who have access to a digital computer.

22. Develop a computer program to simulate the function of an MCU in decoding tab sequential NC blocks. As input, use records of the form

$$\underbrace{t\,001}_{\textbf{n}} \quad \underbrace{t\,07500}_{\textbf{x}} \quad \underbrace{t\,06250}_{\textbf{y}} \quad \underbrace{t\,612}_{\textbf{f}} \quad \underbrace{t\,718}_{\textbf{s}} \quad t \;\; eob$$

All dimension data has four decimal digits, and function data is in *Magic-3* format. Print all data as decimal values.

23. Develop a computer program to simulate the function of a MCU in decoding word address NC blocks. Use the part program of Problem 12. All dimension data has three decimal digits and function data is in *Magic-3* format.

24. Based on the analysis performed in Problems 19 and 20, write a computer routine which, given the following input, will output spindle speed corrections so that surface speed will never vary by more than 1 percent:
   INPUT—start radius, finish radius, horizontal feed, radial feed, and initial spindle speed or required surface speed.
   OUTPUT—a table of spindle speed and time.

Chapter Seven
Numerical Control
Programming

The numerically controlled machine requires information that is initially represented as geometry and process data. It is therefore necessary for a step in the NC process to be occupied by a function that translates conceptual data into information for the machine control unit. This translation process is carried out by NC programming.

A numerical control program is a continuous chain of commands that provides specification of movement and machine activity. The machine control unit of an NC machine can recognize implied data in an NC program, but instructions must be explicitly defined so that no ambiguity or omission exists. The NC program must be coded in a *language* that conforms to the requirements of the MCU. Proper symbolic representation of geometric data and machine commands is essential.

7.1 Manual Programming Methods

The manual preparation of a numerical control program is accomplished by the programmer after consideration of the complex interrelationships between the machine tool and the numerical control unit. These considerations include specific axis nomenclature, tape preparation requirements, and the input requirements of the MCU.

Initially, the programmer describes the movements of the cutting tool with a geometric model, and, based on the resultant calculated coordinates, he specifies control information to provide data that drives the NC machine.

A variety of machine control functions can be specified within a block of numerical control information. The NC program includes specifications of preparatory functions, miscellaneous functions, spindle speed, and feedrate coding. As we have seen, preparatory functions generally serve as indicators to guide the control unit during the issuing of motion and positioning commands. Miscellaneous functions, on the other hand, initiate machine tool operations.

The specification of spindle speed and feedrate coding requires a knowledge of the workpiece machinability characteristics. The feedrate must be specified taking into account the acceleration and deceleration characteristics of the machine tool.

7.1.1 Coordinate System Nomenclature

The standard space coordinate system for NC equipment is the Cartesian reference frame, illustrated in Figure 7.1. Manually developed programs consider movement in the x, y, and z directions with simple angular specifications. Continuous movement in three or more coordinates is invariably accomplished using computer assisted techniques because of the complex calculations involved. Although the Cartesian reference frame is most prevalent, other* reference frames and coordinate nomenclature are available.

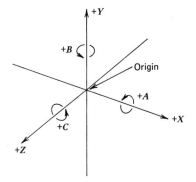

Figure 7.1 Typical axis nomenclature for NC.

7.1.2 An Example of Manually Developed Coordinate Data

During a machining operation with a four-axis machine tool (i.e., x. y, z, and B), the center line of the cutting tool is required to pass through the points (1.500, 2.000, 3.750) and (0.500, 2.000, 3.000) of a stationary Cartesian coordinate system. Using simple geometric relationships, the B-axis value may be determined.

---

* Coordinate systems for various numerically controlled machines can be found in Reference 1.

Referring to Figure 7.2, the value of B is defined as

$$B = \tan^{-1} \frac{\Delta x}{\Delta z}$$

Hence

$$B = \tan^{-1} \frac{(1.500 - 0.500)}{(3.750 - 3.000)}$$
$$= \tan^{-1} (1.383)$$
$$= 53.13°$$

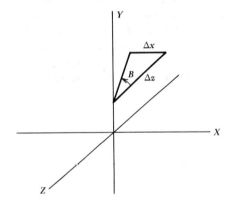

Figure 7.2 B-axis description for stationary 3-D reference frame.

7.1.3 Numerical Control Program Types

There are two different categories of numerical control programming. *Positioning*, also known as *point-to-point* or *straight-cut* programming, is used in NC equipment that locates discrete points without regard to the intermediate path. *Contouring*, also known as *continuous path* programming, is used by NC equipment to describe a continuous locus of points.

The difference between the two techniques is fundamental. Point-to-point methods are used only in applications in which a point is located and then an operation (e.g., drilling) is performed. The path of the cutting tool between points is not defined by the programmer. Therefore, with certain exceptions* the tool cannot be in contact with the workpiece during the movement necessary to travel between points.

Contouring enables the programmer to specify a continuous tool path and is therefore used for continuous cutting operations such as milling.

---

* Some point-to-point machining centers allow straight line cutting along major axes.

7.1.4 An Example of Point-to-Point Operations

Three holes at coordinates $(0.0, 0.0)$, $(0.4, 0.7)$, and $(1.0, 0.4)$ are to be drilled using point-to-point program. Tool movement between points is 4 mpm, and the drilling operation at each point requires 8 sec. For all movement not parallel to one of the axes, the tool travels 1 mm in the x direction for each millimeter along the y axis. The shortest tool path and the total time required to perform this drilling operation may be determined.

The point-to-point tool path is shown in Figure 7.3. To determine the total time, the distance, D, traveled by the tool must be computed. From the geometry

$$d_{11'} = \frac{\Delta x}{\cos 45} = \frac{0.4}{0.707} = 0.566 \text{ m}$$

$$d_{1'2} = 0.3 \text{ m}$$

$$d_{22'} = \frac{\Delta y}{\sin 45} = \frac{0.3}{0.707} = 0.424 \text{ m}$$

$$d_{2'3} = 0.3 \text{ m}$$

$$D = d_{11'} + d_{1'2} + d_{22'} + d_{2'3} = 1.59 \text{ m}$$

Tool travel time $= 1.59 \text{ m}/4 \text{ mpm} \cdot 60 = 23.8$ sec, and drilling time $= 8$ sec/hole $\cdot$ 3 holes $= 24$ sec. Thus, the time required to perform the entire drilling operation is 47.8 sec.

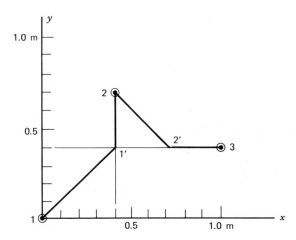

Figure 7.3 Tool path for point-to-point machine.

7.2 Point-to-Point Programming Applications

Normally, positioning machines and point-to-point programming are used to direct a tool to a specific location on a workpiece. Operations such as drilling, tapping, and boring require only a positioning machine, and for these applications the cutting tool is *not* in contact with the workpiece during traverse motion. It is only after the spindle and workpiece attain the programmed position that the machining process begins.

Some point-to-point machines provide a limited continuous path capability. Straight line paths can be milled along either the x or y axis provided there is no change in the z coordinate. Since most positioning controls generate motion simultaneously along x and y axes, a 45° path can also be machined. If cutting occurs during traverse motion, it is important that a programmed feedrate for x-y movement be used.

The ability to machine straight lines using positioning equipment enables the programmer to undertake face milling and pocket milling. For example, if the pocket illustrated in Figure 7.4 were to be machined using point-to-point NC equipment, a straight line tool path could be described as shown in Figure 7.5.

Pocket milling using point-to-point programming techniques requires careful consideration and allowance for cutter diameter and tool path. Points should be chosen so that tool paths overlap, and thereby all excess material can be removed. Unnecessary tool movement should be eliminated (Reference 2).

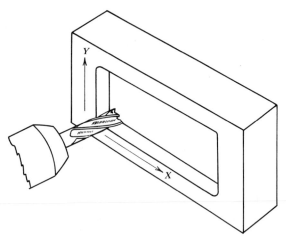

Figure 7.4   Pocket milling.

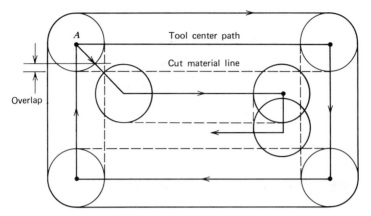

Figure 7.5   Pocket milling tool path. Point *A* is starting location.

7.3 Continuous Path Programming Applications

Continuous path numerical control requires that from two to five axes of a machine be simultaneously and precisely coordinated to follow a predefined locus of points. Often, the cutting tool is in contact with the workpiece as movement occurs. Most continuous path contouring operations involve milling and lathe tools, but continuous path movement is also extensively used for welding, flame cutting, grinding, and other operations such as automatic drafting.

The major difference between point-to-point and continuous path machines is due to interpolation. The control unit for continuous path equipment has the ability to accept coordinate information, interpret it, and then output synchronized movement commands. By precisely controlling the relative movement of axes, a resultant interpolated move can be effected. The calculations necessary to control this process are performed by the MCU, based upon coordinates and feedrates supplied by the programmer.

To follow a uniquely defined path in space, the process reduces the path to line segments. As illustrated in Figure 7.6, straight line *chords* are used to define the curve. As the tolerance dimension, *t*, becomes smaller, the length of each individual chord becomes shorter, and as *t* approaches zero, the number of line segments becomes infinite. Because each line segment must be defined during manual programming, curves requiring small tolerances are tedious to program manually.

Consider the circular arc of length $r\theta$ illustrated in Figure 7.7. Using elementary geometric relations, the number of line segments as a function of tolerance, *t*, can be derived.

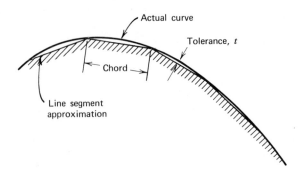

Figure 7.6 Contour approximation using line segments.

Let n be the number of segments; then for an arc subtended by the angle θ, each line segment (chord) would be subtended by the central angle, ϕ, where

$$\phi = \frac{\theta}{n} \qquad (7.1)$$

From plane geometry it can be shown that the chord length

$$l = 2r \sin \frac{\phi}{2}$$

and the normal distance, r_t, from the arc center point to the chord is shown by

$$r_t = \left[r^2 - \left(\frac{2r \sin (\phi/2)}{2} \right)^2 \right]^{1/2} \qquad (7.2)$$

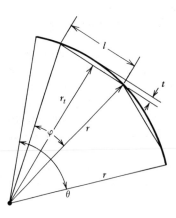

Figure 7.7 Nomenclature for chord tolerance approximation.

Reducing equation (7.2) we obtain

$$r_t = r \cos (\phi/2).$$

The tolerance

$$t = r - r_t = r\left(1 - \cos\frac{\phi}{2}\right)$$

or

$$\phi = 2 \cos^{-1}\left(1 - \frac{t}{r}\right) \tag{7.3}$$

From equation (7.1) we define the number of line segments

$$n = \theta/\phi$$

and from equation (7.3)

$$n = \frac{\theta}{2 \cos^{-1}\left(1 - \dfrac{t}{r}\right)} \tag{7.4}$$

Equation (7.4) yields the functional relationship between the number of line segments needed to approximate a circular arc of length $r\theta$ for a given tolerance, t.

7.3.1 An Example of the Effect of Tolerance on Curve Approximation

A cutting tool must follow a circular path of 0.3 m radius through a 90° angle. The number of linear segments necessary to approximate the curve for tolerances of 0.01 m, 0.001 m, and 0.0001 m may be determined by using equation (7.4) with $\theta = \pi/2$ and $r = 0.3$.

Substituting the appropriate values for t, we obtain:

$$n = 3.03, \ t = 0.01 \text{ m}$$
$$n = 9.61, \ t = 0.001 \text{ m}$$
$$n = 30.4, \ t = 0.0001 \text{ m}$$

Since the number of line segments must be a whole number for the curve to end at the arc's end point, the next highest whole number must be specified. Hence, $n = 4$, 10, and 31, for progressively smaller values of t.

With the number of segments known, the central angle can be recalculated using equation (7.1) and the x and y coordinates developed.

7.3.2 Other Methods of Arc Approximation

Expressions can be developed for tolerance, t, of a given linear approximation of a circular arc when (1) the linear segment is tangent to the arc (tangential approximation) or (2) the linear segment is calculated to be such that it intersects the arc leaving equal tolerance inside and outside the arc (secantial approximation).

Referring to Figure 7.8a, we can write

$$\cos \frac{\alpha}{2} = \frac{R}{R+t}$$

or

$$t = R\left(\sec \frac{\alpha}{2} - 1\right) \tag{7.5}$$

Likewise, from the geometry it can be shown that

$$l = 2R \tan \frac{\alpha}{2} \tag{7.6}$$

Equations (7.5) and (7.6) express tolerance and chord length for tangential approximation of a circular arc.

From Figure 7.8b, we can develop relations for both t and t' for secantial approximation, i.e.,

$$t = R - R_b$$

and expanding

$$t = R - (R + t') \cos \frac{\alpha}{2}$$

$$t = R - R \cos \frac{\alpha}{2} - t' \cos \frac{\alpha}{2}$$

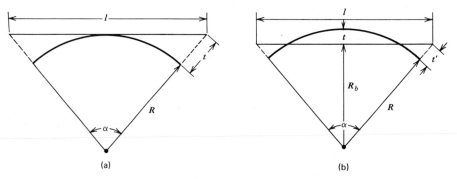

(a)   (b)

Figure 7.8 Tangential and secantial circular arc approximations.

Rearranging,

$$t + t' \cos\frac{\alpha}{2} = R - R \cos\frac{\alpha}{2}$$

Now, since $t = t'$,

$$t = R\left(\frac{1 - \cos\dfrac{\alpha}{2}}{1 + \cos\dfrac{\alpha}{2}}\right) \tag{7.7}$$

The length of the segment, l, is developed from eq. (7.7):

$$\frac{l}{2} = (R + t)\sin\frac{\alpha}{2}$$

Hence

$$l = 2R\left[1 + \left(\frac{1 - \cos\dfrac{\alpha}{2}}{1 + \cos\dfrac{\alpha}{2}}\right)\right]\sin\frac{\alpha}{2} \tag{7.8}$$

7.3.3 Circular Interpolation

The example in Section 7.3.1 illustrates the large number of line segment end points that would have to be calculated and encoded into individual blocks for input to the MCU. To eliminate this requirement, *circular interpolation* has been developed for circular arcs.

Circular interpolation is a function performed by the MCU whereby the motion of the axes is coordinated so that an arc is generated as a quasi-continuous curve. Using the circular interpolation features, only one block of information containing the beginning and end points of the arc, the center point, and the direction of cutter travel (clockwise or counterclockwise) need be specified. For many machine controls, the arc generated consists of short line segments, and the MCU uses the smallest straight line resolution of movement (often 0.004 mm) to describe the arc. Circular interpolation is generally limited to the three principle planes (**XY, XZ, YZ**); however, sophisticated controls which interpolate in all orientations have been developed.

7.3.4 Parabolic Interpolation

Parabolic interpolation extends the simple two point form used for circular arcs and generates a parabolic curve passing through three points in space. It is applied when sculptured patterns or surfaces are required. Whereas linear and circular interpolation lend themselves directly to manual and computer assisted methods, parabolic interpolation requires specialized techniques and equipment.

7.3.5 Cutter Path Offset

To machine a contour, the *cutter path* must be defined. Since the tool has finite dimensions the path must be *offset* such that the cutting surface of the tool is tangent to the workpiece at all times. Hence, to machine the contour shown in Figure 7.9 with a tool of circular cross section of radius, R, a new set of points R units away from *ABCDEF* must be calculated.

For sections *AB, BC, CD*, and *EF* the *cutter offset* is formed by displacing A, B, C, D, E, and F, R units in a horizontal or vertical direction. If applied to all points, however, an offset of R units vertically produces the cut shown in Figure 7.10. The contour required and the actual cut would *not* be the same. The cutting tool must therefore be offset on a locus of points that are *normal* to the workpiece contour at all points.

Accurate offset can be determined only when the normals to a curve, approximated by a series of line segments, have been determined. Consider the situation in which the contour $y = f(x)$ is to be machined using a cutter of diameter $2R$. Initially the programmer chooses a line segment approximation to

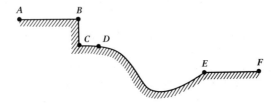

Figure 7.9   Template contour.

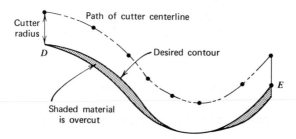

Figure 7.10   Tool path and overcut due to improper tool offset.

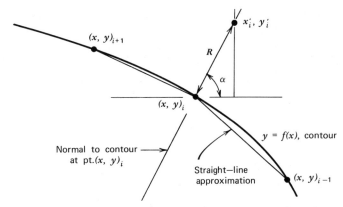

Figure 7.11   Geometry for normal calculation and tool offset from a contour.

satisfy tolerance requirements. At each intersection point the slope $f'(x)$, can be calculated. The slope, N, of the normal to the contour at the point (x_i, y_i) is defined as

$$N = -\frac{1}{f'(x_i)}$$

and it follows that the angle

$$\alpha = \tan^{-1} N$$

defines a line through (x_i, y_i) which represents the locus of all points normal to the contour at that point.

From the geometry shown in Figure 7.11, the offset points (x'_i, y'_i) a distance R from the curve are

$$x'_i = x_i \pm R \cos \alpha$$
$$y'_i = y_i \pm R \sin \alpha$$

(7.9)

The sign convention in equations (7.9) depends upon the direction of offset.

This example, and the ease with which α was calculated, is an oversimplification of the real situation. In practice many contours are not analytic functions and tables of values are *spline* fit to produce smooth contours. These special numerical techniques are carried out with the aid of a computer (Chapter Eight).

7.4 Manual Programming for a Simple Geometry

Manual programming for continuous path applications requires that one *block* of machine control data be developed for each continuous machine movement. Each block consists of *words* which are grouped sets of characters arranged in a prescribed manner. The calculations necessary to develop an NC block were discussed in detail in Chapter Six.

Figure 7.12 illustrates a simple part geometry and the tool path (*ABCDEFGHC*) required to machine it. Initially the programmer defines a reference point or origin from which all other coordinates are displaced.* During machining the tool is moved from the start point, *A*, to a point *B* such that

$$x_B = D_7$$
$$y_B = D_4 - R_c$$

where R_c is the cutter radius. Note that point *B* is offset from contour $C'D'$. Point *C* is offset from $C'H'$ such that

$$x_C = D_5 - R_c$$
$$y_C = D_4 - R_c$$

The contour $C'D'$ can be machined by describing cutter movement *past* a vertical line through D' to point *D*. However, specification of $(x, y)_D$ is not as

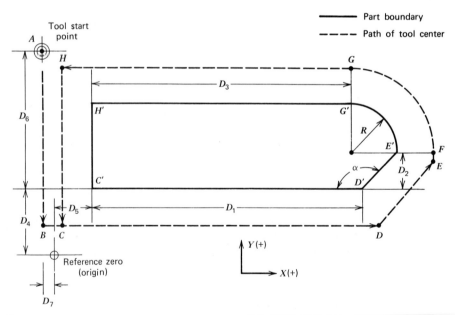

Figure 7.12 Dimensioned part and required tool path for NC part program development.

---

* The reference of the program *zero* to the machine *zero* position is adjustable.

elementary as for points B and C. Point D is the intersection of the offset lines CD and DE. Referring to Figure 7.13,

$$\left. \begin{array}{l} x_D = x_{D'} + R_c \tan \dfrac{\pi - \alpha}{2} \\[2mm] y_D = y_{D'} - R_c \end{array} \right\} \quad \alpha > \dfrac{\pi}{2}$$

and

$$\left. \begin{array}{l} x_D = x_{D'} + R_c \tan \dfrac{\alpha}{2} \\[2mm] y_D = y_{D'} - R_c \end{array} \right\} \quad \alpha \leq \dfrac{\pi}{2}$$

where $x_{D'} = D_1 + D_5$, and $y_{D'} = D_4$.

The contour $D'E'$ is machined by describing cutter path DE (Figure 7.13). The coordinates of point E are:

$$x_E = x_{E'} + R_c \cos\left(\alpha - \frac{\pi}{2}\right)$$

$$y_E = y_{E'} - R_c \sin\left(\alpha - \frac{\pi}{2}\right)$$

where $x_{E'} = D_5 + D_1 + D_2/\tan(\pi - \alpha)$ and $y_{E'} = D_4 + D_2$.

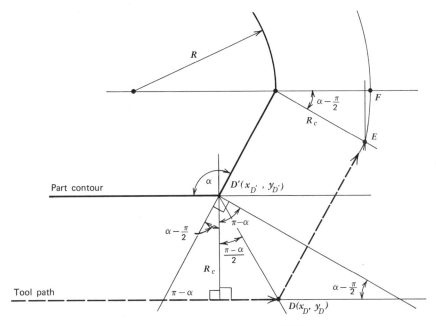

Figure 7.13   Geometry for point D and E calculations.

Contour $E'G'$ is next machined by describing the circular arc EFG. Assuming the NC control unit possesses circular interpolation, the arc may be described without the need to define a straight line approximation. However, since circular interpolation may be specified in only one quadrant, $E'G'$ must be machined by describing two arcs, EF and FG. Referring to Figure 7.14, circular interpolation requires that the NC programmer supply incremental coordinate information to define the initial arc point, Δi and Δj (in two dimensions), and the final arc point, Δx, Δy. For arc EF, the incremental values are seen to be

$$\Delta i = R + R_c \cos \left(\alpha - \frac{\pi}{2} \right)$$

$$\Delta j = -R_c \sin \left(\alpha - \frac{\pi}{2} \right)$$

and

$$\Delta x = R_c \left[1 - \cos \left(\alpha - \frac{\pi}{2} \right) \right]$$

$$\Delta y = R_c \sin \left(\alpha - \frac{\pi}{2} \right)$$

where R is the part radius and R_c is the cutting tool radius. Since arc FG travels a full 90°, the specification of the Δi, Δj, Δx, and Δy coordinates is greatly simplified:

$$\Delta l = R + R_c$$
$$\Delta j = 0$$
$$\Delta x = -(R + R_c)$$
$$\Delta y = (R + R_c)$$

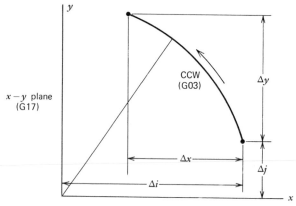

Figure 7.14 Two-axis circular interpolation.

Finally, cutter path GHC is developed. Point H is equidistant from lines $H'G'$ and $H'C'$, hence

$$x_H = D_5 - R_c$$
$$y_H = D_4 + D_2 + R + R_c$$

Although this completes the specification of dimensional information, NC blocks also contain nondimensional words. Each block requires a sequence number and suitable preparatory function for each tool movement. The feed function must be calculated and specified, and miscellaneous functions must be determined.

Using the information presented in Chapter Six and Figure 7.12, Table 7.1 shows the completed NC program for the following dimensional information. Referring again to Figure 7.12,

$D_1 = 3.5$	$D_2 = 0.5$	$D_3 = 3.25$
$D_4 = 1.0$	$D_5 = 0.5$	$D_6 = 2.5$
$D_7 = 0.25$	$\alpha = 135°$	$R = 0.75$

The cutter radius $R_c = 0.5$.

Table 7.1
NC Program for Simple Part Geometry

Word address characters[a]	Comments
n001 g08 x − 2500 y10000 z0 f518 s716 t01 m03 *eob*	A
	to
n002 g09 y5000 f430 m08 *eob*	B
n003 g01 x42071 *eob*	B to D
n004 x53535 y11464 *eob*	D to E
n005 g17 *eob*	Define plane
n006 g03 x1464 y3535 i11035 j3535 f420 *eob*	E to F
n007 g03 x − 12500 y12500 i12500 j0 f425 *eob*	F to G
n008 g01 x5000 y27500 f430 *eob*	G to H
n009 g09 y5000 *eob*	H to C
n010 m05 *eob*	Spindle stop
n011 m09 *eob*	Coolant stop
n012 m00 *eob*	Stop

[a] Each block begins with an **n** word and ends with *eob*. No spaces are punched between characters.

7.5 Computer Assisted NC Programming

The programming example presented in the previous section shows that even the simplest component requires a specific series of calculations to generate dimensional data so that NC blocks can be generated. As component design becomes more complex, the number of calculations increases substantially, and the manual development of programs becomes an error prone, time consuming task. To overcome these problems, the digital computer is used.

Computer assisted numerical-control programming provides a number of significant benefits. The speed and accuracy of modern computing systems enable the elimination of many manual tasks. However, a more important benefit is the generality of NC programming languages and the application of logical operations to provide more efficient NC programs. Computerized methods enable the NC programmer to describe cutter paths that could not be specified using manual techniques.

The general information flow in an NC programming system is illustrated schematically in Figure 7.15. The computer serves as the interface between the component design and tool path description which comprises the part program and the NC cutter location data as specified by the input media.

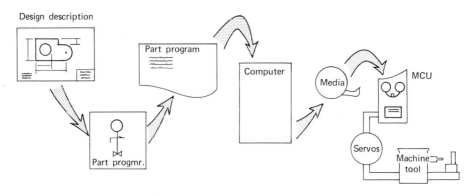

Figure 7.15 Schematic of computerized NC programming information flow.

7.5.1 The Computerized Part Program

The *part program* is a series of coded statements whose format and order are developed in the form of a language. For NC programming applications the language contains a geometry capability and a method through which tool motion and machine commands are described.

A computer program *processes* the part program. The computer program, or processor, uses the syntactical rules of the language in which the part program was written to translate coded language statements into a tool path description.

7.5.2 The Processor

The processor is comprised of elements illustrated in Figure 7.16. The computer program contains a number of processing subsections, each of which performs a specific operation on the input part program. The initial analysis subsection of the processor is called the *translator*. The translator transforms the part program language into executable computer instructions that are processed by the second subsection, the *arithmetic element*. All of the mathematical analysis used to develop cutter location coordinates is performed in this subsection.

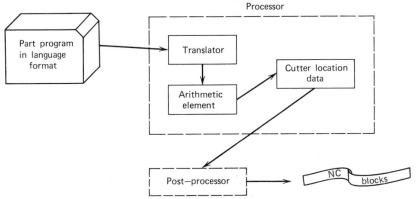

Figure 7.16 A part programming system—processor elements and post-processor.

The cutter location coordinates, as they are output from the arithmetic section, are in a generalized form and must undergo further processing. The next phase of the computer interface generates the dimension and nondimension information necessary to develop an NC block.

7.5.3 The Postprocessor

The processor is a generalized program concerned with developing cutter location data based on geometry and motion information. A *postprocessor* is a computer program which is written to develop all of the NC block information for a *specific* MCU. Cutter location coordinates (called CL data) are input to the postprocessor, and NC blocks are output. Postprocessor output may require conversion of cards or magnetic tape to punched tape.

Because a postprocessor outputs data specific to a given control unit, a typical computerized NC programming system has a different postprocessor program for each MCU in use.

7.6 A Simple Part Programming Language—SPPL

A computer assisted part program is written using a series of statements whose sequence, construction, and content are specified by a language containing precise grammatical rules. All computer languages contain a vocabulary that enables the programmer to communicate with the machine. The vocabulary and supplementary symbolic information for a program are translated (sometimes, the term *compiled* is used) into machine language that the computer can execute.

For NC applications, the programmer requires a language that has:

1. A geometry description capability that enables him to describe necessary calculations without having to execute them.
2. A method to describe tool motion.
3. A means for specifying machine tool information such as feeds, speeds, and miscellaneous functions.

These capabilities are available in many different NC programming languages developed for specific application areas such as turning or two-axis contouring. For general NC programming applications, powerful languages with extensive three-dimensional geometric capabilities are also available. The most widely used NC programming language is APT.*

An NC programming language is *syntax directed.* That is, the language is governed by a set of rules that dictate how the vocabulary elements are ordered to form statements. Syntax refers to the language structure and is derived from a study of natural languages. The sentence "A man climbs the hill" may be diagrammed as shown in Figure 7.17. The diagram displays the syntax of the sentence in schematic fashion.

Consider the development of a typical NC programming language recalling that, like the English language, it should be syntax directed. Specifically, the language, which we shall call SPPL (*simple part programming language*), considers only point, line, and circle geometries in two dimensions; tool motion

---

* *Automatically Programmed Tool*, to be discussed in a later section.

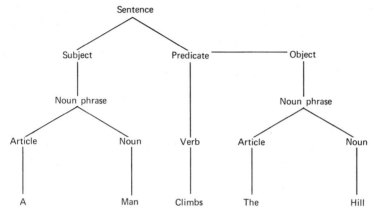

Figure 7.17   A sentence diagram.

relative to a given location; and special statements which define the postprocessor, specify cutter diameter and tolerance information, and spindle/coolant data. Geometric entities are defined by direct specification or by intersection, tangency, or angular definitions. By considering the structure of SPPL, an understanding of the mechanics of many actual NC programming languages can be achieved.

SPPL statements consist of vocabulary, actual numbers, symbolic names for previously defined entities, and punctuation. Symbolic names are used to uniquely define a given geometry entity for later use by the programmer. Three basic statements are considered: (1) *geometry statements*, (2) *motion statements*, and (3) *special statements*. Each statement consists of *major* words that define a specific geometry or operation, and *minor* words that modify information within the statement.

Let us develop a syntax for SPPL. Three types of statements exist in our language. The statement structure for each type is defined here:

GEOMETRY STATEMENT: symbolic name = major word/other symbolics, numbers, and minor words.

MOTION STATEMENT: major word/other symbolics, numbers, and minor words.

SPECIAL STATEMENT: major word/minor word, and numbers.

Hence, the general form of an SPPL statement can be illustrated as follows:

GOTO/WORK, NEXT, FRIDAY, AT, EIGHTPM

This sentence, which is not a legitimate SPPL statement, does exhibit all the properties of a motion statement. The major word GOTO is self explanatory. Symbol names WORK, FRIDAY, and EIGHTPM represent known information, and minor words NEXT and AT qualify this data. If, for example, other minor words, e.g., THIS instead of NEXT, and BEFORE instead of AT, are substituted,

GOTO/WORK, THIS, FRIDAY, BEFORE, EIGHTPM

the entire meaning of the statement is modified.

The vocabulary for SPPL is outlined in Table 7.2. The table indicates the major and minor words that can be associated with each type of statement. Hence, we can develop a formal syntax for SPPL.

Table 7.2
Vocabulary for SPPL—Major and Minor Words

Sentence type	Major words	Minor words	Description
Geometry	POINT		Point defined by x, y[a]
	LINE		Line defined by x_1y_1, x_2y_2[a]
	CIRCLE		Circle defined by x_c, y_c, R[a]
		TANTO	Tangent to a given element
		INTOF	Intersection with given element
		ATANGL	At angle specified
		LEFT/RIGHT	Circle tangency qualifier
Motion	GFWD		Go forward
	GRIGHT		Go right
	GLEFT		Go left
	FROM		From a given location
		ON	On
		PAST	Past
		TO	To
Special	POSTPR		Postprocessor type
	TOLER		Tolerance specification
	SPINDLE		Spindle speed
		CW	Clockwise
		CCW	Counterclockwise
	COOLANT		Coolant
		ON	On
		OFF	Off
	CUTTER		
	STOP		

[a] Indicates *specific* definition; other definitions using minor words are available.

As an example, consider a geometry statement that defines a line, L1, which passes through a given point, PT1, and is tangent to a predefined circle, CIRCLE1.

$$L1 = LINE/PT1,\ RIGHT,\ TANTO,\ CIRCLE1$$

This statement conforms to the formal syntax definition, and therefore can be correctly translated by the SPPL processor.

7.6.1 An Example of SPPL Geometry Statements

Considering the SPPL geometry statements listed, draw the geometry described by the code and label appropriate entities.

$$PT1 = POINT/0.0, 0.0$$
$$C1 = CIRCLE/0.0, 1.0, 1.0$$
$$L1 = LINE/0.0, 0.0, 4.0, 0.0$$
$$PT2 = POINT/4.0, 1.0$$
$$L2 = LINE/PT2, ATANGL, 45.0$$
$$L3 = LINE/PT2, RIGHT, TANTO, C1$$

The enclosed geometry shown in Figure 7.18 is described by the above SPPL statements. Given this geometry, SPPL may be used to define the point PTA. Using indirect specification

$$PTA = POINT/INTOF, L1, L2$$

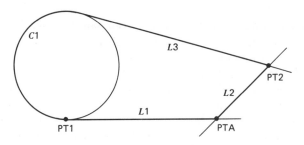

Figure 7.18 Part geometry for Example.

7.6.2 SPPL Program Translation

The SPPL processor program translates the programming statements and develops the mathematical relationships necessary to describe geometry. Given the statement

$$\text{LINE1} = \text{LINE/PTA, LEFT, TANTO, CIRCLE1}$$

the steps performed by the SPPL translator and arithmetic element are outlined below.

The translator decodes the statement by (1) resolving symbolic references to actual values, and (2) setting up branching to correct arithmetic routines. This process, sometimes called *parsing* the statement, is performed after the statement syntax is verified. Each symbol is isolated and appropriate values retrieved from storage locations previously allocated for the symbol. Hence,

$$\text{PTA} \leftrightarrow x_A, y_A$$
$$\text{CIRCLE1} \leftrightarrow x_1, y_1, R_1$$

The major word, LINE, is recognized along with an analysis request for tangency to a circle through a given point. All statements are translated before the arithmetic element is invoked.

In the arithmetic element, a line can be defined by the general equation

$$Ax + By - D = 0,$$

and a circle by

$$(x - x_c)^2 + (y - y_c)^2 = r^2$$

By substituting x_A, y_A and x_1, y_1 and R_1 into these equations, quadratic solutions for the ratios $-B/A$ and D/A can be developed. The choice of values is dictated by the minor word LEFT.

7.6.3 Motion and Special Statement Commands

Once a part has been defined with geometry statements, tool movement is specified using motion statements. These commands, indicating *right*, *left*, and *forward*, are specified in relation to the tool moving along the surface. This directional sense is illustrated in Figure 7.19.

Tool motion must be limited by the NC programmer. Therefore, once direction has been specified, instructions indicate at what point the tool must change direction or stop. In an SPPL context, the minor words ON, PAST, and TO control the end motion along a particular segment of the cutter path.

Figure 7.20 illustrates how the motion statement is used to control tool movement. Lines L1, L2, and L3 have been previously defined using geometry

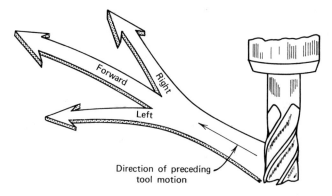

Figure 7.19 Direct specification from the tool viewpoint.

statements. The cutter must move along lines L1, L2, and L3, the motion terminating at point *B*. The proper motion commands to describe this path are

GFWD/L1, PAST, L2

GLEFT/L2, TO, L3

GRIGHT/L3, ON, L4

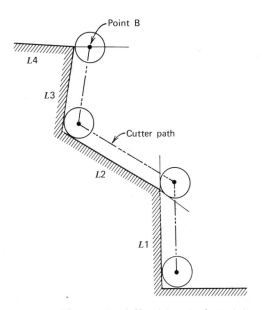

Figure 7.20 Tool path defined by motion statements.

The minor words indicate the end position of the cutter for each path segment. The cutter travels in the specified direction until it is tangent (in the case of PAST or TO) to the indicated line or until its center line rests on the line (ON). PAST indicates that the second tangency point is to be used, whereas TO implies the first point.

Special statements are used to (1) pass control to the processor—for example, the POSTPR statement indicates the name of the postprocessor to which the cutter location data is to be sent; (2) indicate mathematical information such as tolerances for linear approximation of curves; and (3) specify miscellaneous functions that must be included when the cutter location data is postprocessed. For example, the statement

COOLANT/ON

would cause the processor to create an indicator so that the postprocessor would issue an **m07** or **m08** command. Likewise,

SPINDLE/ON, CCW

would cause the postprocessor to issue an **m04** command. (See Table 6.4.)

7.6.4 An SPPL Program

We can see that SPPL and, in a more general sense, all NC programming languages are somewhat conversational in nature. Furthermore, by eliminating the tedious calculation of tolerance, tangency and intersection points, and feed and speed functions, the numerical control programming task is greatly simplified.

Reconsider the simple part (Figure 7.12) for which a manual NC program was developed. Using the values specified for D_1 through D_7, R, R_c, and α in the text, we can develop an SPPL program for the part.

Using the dimensions illustrated in Figure 7.21 all values are specified with respect to the origin (0., 0.).

PTA = POINT/−0.25, 3.5

PTC = POINT/0.5, 1.0

PTH = POINT/0.5, 2.25

PTCR = POINT/3.75, 1.5

PTD = POINT/4.0, 1.0

The lines comprising the boundary of the part can be described using end points only. However, this approach is not used in all cases, and to illustrate the

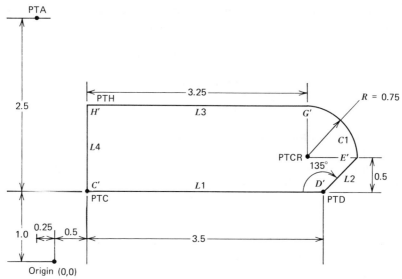

Figure 7.21 Dimensioned part geometry for SPPL example.

use of minor words, we can write

$$L1 = LINE/PTC, \; PTD$$
$$L2 = LINE/PTD, \; ATANGL, \; 45.0$$

Since line L3 depends upon the definition for circle C1,

$$C1 = CIRCLE/PTCR, \; 0.75$$

Then,

$$L3 = LINE/PTH, \; LEFT, \; TANTO, \; C1$$
$$L4 = LINE/PTC, \; PTH$$

A point, PTE, at which line L2 and circle C1 intersect is defined as

$$PTE = POINT/INTOF, \; L2, \; C1, \; YLARGE$$

It should be noted that a new minor word, YLARGE, has been introduced. YLARGE qualifies the solution by indicating that the intersection point with the largest y value should be chosen.

As we shall see in our discussion of the motion statements, an additional check line, LCHK, must be described to aid in proper specification of cutter motion in the vicinity of point E.

$$LCHK = LINE/PTCR, \; PTE$$

The geometry description of the part is now complete.

The cutter motion begins at point PTA and continues as illustrated in Figure 7.12. Referring to the Figure, path AB is described

GFWD/FROM, PTA, PAST, L1

Recalling the cutter's sense of direction, path *BCD* is described as

GLEFT/L1, PAST, L2

which indicates that the cutter is to move along a path parallel to line L1 until it has passed and is tangent to L2.

To illustrate the use of the check line, Figure 7.22 shows the cutter path along L2 and C1. Path P_1P_2 is described by

GLEFT/L2, ON, LCHK

Note, however, that at P_2 the cutter is *not* tangent to the circle C1 and is not in contact with the component periphery. It must be moved to a point of tangency

GLEFT/LCHK, TO, C1

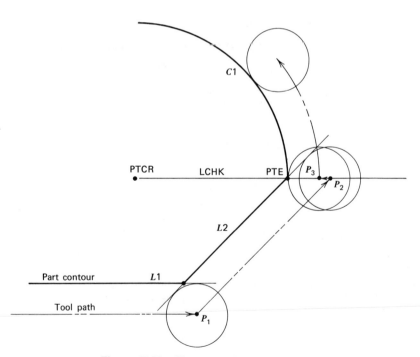

Figure 7.22 The use of a *check line*.

At this location, the cutter center is at P_3. Motion around the arc is defined in a single statement

GRIGHT/C1, PAST, L4

the remaining commands are

GFWD/L3, PAST, L4

GLEFT/L4, PAST, L1

Since the cutter path has been completely defined the final requirement is to specify any special statements that may be necessary. Cutter and tolerance

Table 7.3
An SPPL Program

PTA = POINT/ − 0.25, 3.5
PTC = POINT/0.5, 1.0
PTH = POINT/0.5, 2.25
PTCR = POINT/3.75, 1.5
PTD = POINT/4.0, 1.0
L1   = LINE/PTC, PTD
L2   = LINE/PTD, ATANGL, 45.0
C1   = CIRCLE/PTCR, 0.75
L3   = LINE/PTH, LEFT, TANTO, C1
L4   = LINE/PTC, PTH
PTE   = POINT/INTOF, L2, C1, YLARGE
LCHK = LINE/PTCR, PTE
CUTTER/0.500
TOLER/0.001
COOLANT/ON
SPINDLE/ON, CCW
GFWD/FROM, PTA, PAST, L1
GLEFT/L1, PAST, L2
GLEFT/L2, ON, LCHK
GLEFT/LCHK, TO, C1
GRIGHT/C1, PAST, L4
GFWD/L3, PAST, L4
GLEFT/L4, PAST, L1
SPINDLE/OFF
COOLANT/OFF
POSTPR/POSTPNAME[a]
STOP

[a]Where POSTPNAME is the name of a postprocessor for a specific machine tool.

information are specified with

CUTTER/0.500

TOLER/0.001

The postprocessor, spindle, and coolant commands must also be supplied. By developing a new major word, FEEDRT, the tool feedrate can also be specified. The complete SPPL program for the part illustrated is listed in Table 7.3.

7.7 The Postprocessor

A postprocessor is a computer program called by the processor and used to convert cutter location data into a medium that is understandable by an MCU. Since the conversion process performed by the postprocessor varies in complexity in direct proportion to the sophistication and capability of the machine tool, it follows that simple postprocessors are needed for point-to-point machines, whereas more complex programs are required for multiaxis contouring machines. A typical postprocessor contains five elements: *input*, *motion analysis*, *auxiliary functions*, *output*, and *control and diagnostics*.

The *input element* reads the cutter location data and miscellaneous information that is output from the processor. It verifies the format of the data and transfers appropriate values to other elements of the postprocessor.

The *motion analysis element* contains the dynamics and geometry sections. The geometry section performs coordinate transformations to convert the general CL data into specific machine tool coordinates. When rotary axes are available, the CL data also contains direction cosines that must be translated into rotation commands for the MCU. In addition to these transformations, the geometry section insures that the machine's physical limits are not exceeded and that the tool does not cut into any part of the machine. Finally, it is the job of the geometry section to select proper linear and rotary motions and to insure that the resultant path is within tolerance.

The dynamics section of the motion analysis element calculates the appropriate tool velocity based on servo type and the acceleration/deceleration characteristics of the machine tool. It also serves to optimize the acceleration/deceleration distance so that the tool will travel at the programmed feedrate for the greatest possible distance along a segment.

Figure 7.23a depicts a typical deceleration curve for a machine tool control unit with exponential feedrate control built into the MCU interpolator. From the curve, the relationships which are used by the postprocessor to calculate output feedrate can be developed (Reference 3). The velocity of the tool at time, t, is

$$v = v_F + (v_i - v_F) e^{-t/\tau} \qquad (7.10)$$

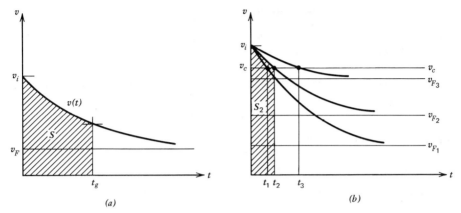

Figure 7.23 (a) Typical exponential deceleration curve. (b) Selection of v_F for a given cornering feedrate, v_c.

and taking the logarithm of both sides, the time

$$t = \tau \ln \left(\frac{v_i - v_F}{v - v_F} \right) \tag{7.11}$$

where v_i is the initial feedrate for a given segment, v_F is the output feedrate to the control unit, τ is the MCU time constant, t is the machining time ($t = 0$ when $v = v_1$), and v is the feedrate of the tool at time, t.

As the tool slows down, the feedrate, v, can be determined from the curve at any given time. At a given time, t_g, the distance, s, traveled by the tool is calculated by integrating equation (7.10),

$$s = \int_0^{t_g} v_F \, dt + \int_0^{t_g} (v_i - v_F) \, e^{-t/\tau} \, dt$$
$$s = v_F t_g + \tau (v_i - v_F) \, e^{-t_g/\tau} \tag{7.12}$$

The shaded region in Figure 7.23a represents the distance, s.

The postprocessor can control the time and distance in which a computed cornering feedrate, v_c, is achieved by increasing or decreasing the value of the input feedrate, v_F. Since v_F is the asymptote of the deceleration curve, the steepness of the curve is varied, and t and s change accordingly. By iterating on the value of v_F (illustrated in Figure 7.23b), optimum deceleration conditions are obtained. It should be noted that the postprocessor selects v_F within the constraints defined by the machine tool and control unit. Optimum acceleration characteristics are obtained in a like manner.

7.7.1 Calculation of Deceleration Time

For a given machine tool, it is necessary to reach a specific deceleration value so that cornering error is kept to a minimum. The tool travels at a constant feedrate, v_i, and must decelerate to a control block feedrate of v_F. A general relation for the time required to reach a specific deceleration can be developed.

From equation (7.10)

$$v(t) = v_F + (v_i - v_F)\, e^{-t/\tau}$$

The acceleration/deceleration:

$$\dot{v} = \frac{dv}{dt} = -\frac{1}{\tau}(v_i - v_F)\, e^{-t/\tau}$$

Rearranging, and taking the natural logarithm of both sides

$$\ln\left(\frac{\dot{v}}{v_F - v_i}\right) = \frac{t}{\tau}$$

Solving for t,

$$t = \tau \ln\left(\frac{v_F - v_i}{\tau \dot{v}}\right) \qquad\qquad (7.13)$$

the deceleration time to slow from v_i to v_F.

7.7.2 Auxiliary Postprocessor Functions

An additional function of the motion analysis element is the generation of NC blocks for circular and parabolic interpolation. A typical part programming language, such as SPPL, provides for circular cutter path without regard to quadrant. The postprocessor takes the *canonical* form of the circle, defined in a given plane, and generates circular interpolation instructions. A basis for this analysis is that no quadrant boundary may be crossed in a single NC block. Because the postprocessor performs necessary segmentation, the NC programmer need not worry about the quadrant restriction.

The *auxiliary function element* provides for the output of miscellaneous and preparatory command codes. For example, the SPPL program contains the statements

$$\vdots$$

COOLANT/ON

SPINDLE/ON, CCW

$$\vdots$$

SPINDLE/OFF

COOLANT/OFF

$$\vdots$$

SEQ	CARDNO	G	X	Y	Z	TABLE FRND	SC	M	I-R	ARC CENTERS	TOOL	TIME	FEDRAT	SPINDL
1	ORIGIN		.0000	10.0000	15.0000							0.0	0.3	0.
2								79				0.0		
3		*						06				0.0		
4												0.0		
5		00	4.0000	7.0000	9.0000			07				0.0		
6						.000	6300	03				0.0	1.0	300.
7		17										0.0	1.0	300.
8		01	3.3833	5.9500	4.9500	4100						1.22	1.0	300.

THE FOLLOWING COMMANDS WILL FINISH MILL CONTOUR

SEQ	CARDNO	G	X	Y	Z	TABLE FRND	SC	M	I-R	ARC CENTERS	TOOL	TIME	FEDRAT	SPINDL
9			1.6310	5.0303								3.19	1.0	300.
10		02	1.5853	5.0353					1.6033	5.0926		3.24	1.0	300.
11		03	−5.2474	.5759					.0000	.0000		10.07	1.0	300.
12		02	−5.3596	.6124					−5.3070	.5934		10.18	1.0	300.
13		01	6.6310	1.6932	4.9500							12.82	1.0	300.
14					9.9500							17.82	1.0	300.
15												26.17	1.0	300.
16			−4.2136	6.2940	4.9500							31.17	1.0	300.
17		02	−1.7207	5.0101								33.57	1.0	300.
18		02	−1.6741	5.0065					−1.6932	5.0634		34.02	1.0	300.
19		03	3.7467	3.7189					−.0000	.3000		42.75	1.0	300.
20		02	3.7616	3.8144					3.7893	3.7611		42.84	1.0	300.
21		01	5.1334	4.5280	4.9500							44.39	1.0	300.
22					9.9500			06				49.39		
23								09				49.39		
24							0000	05				49.39		
25								00				49.39		
26														

STCP

REMOVE CLAMPS FROM THE 11,1 AND 3 O&CLOCK POSITIONS ON FIXTURE

THE FOLLOWING COMMANDS WILL SPOT DRILL 50 HOLES IN LARGE B-C.

SEQ	CARDNO	G	X	Y	Z	TABLE FRND	SC	M	I-R	ARC CENTERS	TOOL	TIME	FEDRAT	SPINDL
27		00	7.6000	.0030	4.0000		6800	07				49.39	1.0	800.
28			7.5401	.9525				03				49.42	1.0	800.
29			7.3612	1.8930								49.42	1.0	800.
30			7.0663	2.7977								49.43	1.0	800.
31			6.6599	3.6613								49.43	1.0	800.
32			6.1485	4.4672								49.44	1.0	800.
33			5.5402	5.2026								49.44	1.0	800.
34			4.8444	5.8559								49.44	1.0	800.
35			4.0723	6.4169								49.45	1.0	800.
36			3.2359	6.8767								49.45	1.0	800.
37			2.3485	7.2280								49.46	1.0	800.
38			1.4241	7.4654								49.46	1.0	800.
39			.4772	7.5850								49.47	1.0	800.
40			.4772	7.4654								49.47	1.0	800.
41			−1.4241	7.4654								49.47	1.0	800.
42			−2.3485	7.2290								49.48	1.0	800.

Figure 7.24 Sample computer printout from a Pratt and Whitney Machining Center postprocessor (*courtesy Avco-Lycoming Division*).

The SPPL processor passes these statements together with geometric data as part of a *CL file*. The postprocessor's auxiliary function element translates the machine control commands as follows:

The COOLANT/ON statement translates to an **m08** command. The SPINDLE/ON, CCW becomes an **m04** block, and the COOLANT/OFF and SPINDLE/OFF statements become **m09** and **m05** commands, respectively.

Many postprocessors support a special CYCLE capability. A single statement of an NC programming language triggers a command in the postprocessor that enables the machine tool to perform an internally preprogrammed operation. In this way, many machine tool movements are implied by a single statement. For example, the special cycle for drilling might be

CYCLE/DRILL, R, *pt*, F, *depth*, IPM, *feedrate*

where the major word is CYCLE and the minor word DRILL specifies the cycle type. The other minor words R, F, and IPM indicate the point of termination for rapid approach, feed into the workpiece, and feedrate, respectively. When the postprocessor recognizes a CYCLE command, it generates the appropriate preparatory function (e.g., **g81**) that triggers internally preprogrammed movement.

The *output element* of the postprocessor generates two types of output: (1) the actual numerical control blocks in a media form that can be either directly input to the MCU or easily converted into a form for direct input to the MCU, and (2) computer printout, illustrated in Figure 7.24. The printout from the postprocessor represents each NC block in a readable format. The printout illustrated in the figure was generated using a Pratt and Whitney Machining Center postprocessor for the APT language. Comments are carried through from the APT statement text.

The *control and diagnostic element* of the postprocessor is necessary to insure that a proper flow of information occurs in the program and that analysis errors are diagnosed and brought to the attention of the NC programmer.

7.8 Processor Languages in Industry—APT

As the number and complexity of NC programming applications have grown, many NC languages have been developed. Most processors, and the part programming languages which comprise their input, are written for specific NC applications on a given machine tool. Machine tool builders and computer manufacturers have prepared processors for their own equipment, and certain manufacturing users have written special purpose programs for their own unique applications (References 4 and 5). However, a number of general purpose processors have been written for use throughout the industry.

The most comprehensive and best known of all NC language processors is APT (*automatically programmed tool*). APT provides the generality and computing power necessary to solve complex NC problems.

In the early days of numerical control (1954–1958), each company had to develop its own NC program processor. It was soon decided that a single language processor capable of handling a broad cross section of NC tasks was required. Therefore, the members of the Aerospace Industries Association pooled their talent and resources and began the development of a new processor language as a cooperative project. This language, called APT, is under continual development to provide ever more sophisticated capabilities.

Like the programming language, SPPL, developed in this text, APT contains English-like major and minor words that have specific meanings. The statement structure and the language concept of APT are similar to SPPL. The APT processor program supports a language of more than three hundred words. The vocabulary has been designed to be open-ended, so that new words representing new capabilities can be incorporated into the language. Different versions of APT have been written for use on different computers, with less powerful versions available for smaller machines.

The APT program processor and the APT language have been designed to provide maximum flexibility for the NC programmer. As with any general processor, the major functions of the APT system are:

1. Geometric definitions.
2. Tool definition and motion statements.
3. Machine tool functions.
4. Computer system commands.

The APT language statements are put together in a sequence that describes the geometry, moves the tool along the surface described, activates various machine tool functions, and makes requests of the computer system.

The APT language provides for the following geometric elements: points; point sets or *patterns*; lines; planes; vectors; circles; circular cylinders; conic cylinders; spheres; cones; quadrics (ellipsoid, paraboloid, etc.); *ruled* surfaces (surfaces generated by straight lines, called rulings, joining corresponding points on two space curves); tabulated cylinders (splined curves); matrices (used for coordinate transformation); and in some versions of APT, a *sculp-tured surface* capability.

Motion commands in APT are specified for absolute or incremental movement. The commands

GOTO/(absolute position)

GODLTA/(incremental move)

refer to these kinds of movements. Not only is the direction of travel of the tool controlled using GO (*up, down, left, right, back, forward*) commands, but the orientation of the tool with respect to the *drive surface* (see Figure 7.25) can be specified. The commands TLLFT, TLRGT, and TLON indicate the position of the tool axis with respect to the drive surface. Referring again to Figure 7.25, the *drive surface* and *check surface* are planes containing the lines L_1 and L_2, respectively, and parallel to the tool axis. The *part surface* is the plane containing *both* L_1 and L_2. For the figure shown, the APT motion commands are

GOFWD/TLRGT, L1, TO, L2

GORGT/TLRGT, L2, . . .

Machine tool commands are specified in APT, but are interpreted and coded by the appropriate postprocessor. Except for spelling differences, the discussion of machine tool commands for SPPL can be applied to APT.

APT contains a number of language capabilities that enhance its overall computing power. These features provide for a subprogram or MACRO facility, arithmetic statements, and looping logic that are analogous in many ways to those available in an all-purpose programming language such as FORTRAN or PL/1.

The MACRO feature allows the NC programmer to preprogram repetitive operations with symbolic parameters instead of actual values. Then, at the point in the APT program at which the operation is required, the MACRO is

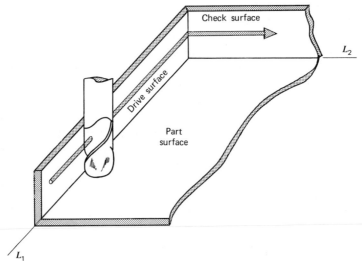

Figure 7.25 Drive, check, and part machining surfaces for APT.

called with actual data substituted for those parameters that were symbolically coded. For example, assume that it is necessary to mill a rectangular path whose sides vary in length. Cutter diameter is also a variable. We define the MACRO, BOX, as

> BOX = MACRO/DIAM, INCRX, INCRY
> CUTTER/DIAM
> GODLTA/INCRX, 0.0, 0.0
> GODLTA/0.0, INCRY, 0.0
> GODLTA/(− INCRX), 0.0, 0.0
> GODLTA/0.0, (− INCRY), 0.0
> TERMAC

Whenever the rectangular cut is desired, motion statements are used to position the cutting tool at the lower left hand corner of the *box* ; then a statement in the following form must be coded

> CALL/BOX, DIAM = *cutter diameter*, INCRX = *x-value*, INCRY = *y-value*

If a 0.2-m × 0.3-m path must be machined using a 12-mm diameter tool, APT statement would be

> CALL/BOX, DIAM = 0.012, INCRX = 0.2, INCRY = 0.3

By using the MACRO facility a single CALL statement replaces five APT commands. When an operation is required in numerous locations, a large number of statements are therefore eliminated.

In many cases the design drawing does not contain the values necessary to generate APT commands. The arithmetic computation feature eliminates many manual calculations that would have to be performed before programming began. For example, to calculate the length of the hypotenuse of a right-angled triangle, with base, B, and height, H, known, the following statement could be written in an APT program:

> HYPOT = SQRTF (B**2 + H**2).

Readers familiar with FORTRAN will recognize that the APT notation is the same. All basic operations (+, −, *, **, /) and functions (sine, cosine, tangent, square root, etc.) are available.

Arithmetic logic may be used within an APT command statement, for example,

> CUTTER/(B*A + 3.0 − D)

where B, A, and D have all been previously defined. Hence, the arithmetic statement provides an added computational feature to the language.

Looping logic is used to alter the sequential flow of the NC part program by repeating a given set of APT statements based on a logical condition. A loop is generally used when the same operation is performed with minor modification in a repeated sequence. For example, consider the MACRO BOX developed to cut a rectangular path. Assume that a rectangular path 0.072 m wide must be machined using a 12-mm diameter cutter. The outside perimeter is 0.4 m × 0.25 m, and the origin (0.0, 0.0, 0.0) is at the lower left hand corner of the rectangle, as illustrated in Figure 7.26. Using looping logic and arithmetic statements the operation is defined as

```
            D = 0.012
            XS = 0.006
            YS = 0.006
            LX = 0.4 − D
            LY = 0.25 − D
            LOOPST
BEGIN)      GODLTA/XS, YS, 0.0
            CALL/BOX, DIAM = D, INCRX = LX, INCRY = LY
            XS = XS + D
            YS = YS + D
            LX = LX − D
            LY = LY − D
            IF (LX − 0.256) STOP, STOP, BEGIN
STOP)       LOOPND
              ⋮
```

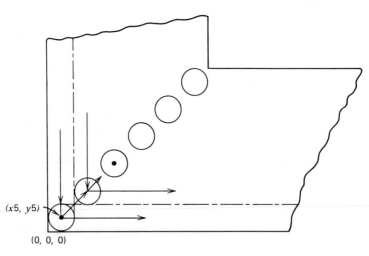

Figure 7.26 Lower left-hand corner of machining path for APT macro BOX.

The APT statements listed above define the six rectangular paths necessary to machine the 0.072 m wide path. The arithmetic statements within the loop relieve the NC programmer of the task of calculating the starting point (XS, YS) and lengths (LX, LY) of the cutter paths. The logical IF statement terminates the loop (and the machining operation) when the base of the rectangle to be cut is 0.256 m or less. It should be noted that in the interest of clarity, cutter overlap was not considered.

References

1. *Machine Axis and Motion Nomenclature*, National Aerospace Standard 938, AIA publication.
2. Childs, J. J., *Principles of Numerical Control*, The Industrial Press, New York, 1965, pp. 59–62.
3. Bobrowicz, V., "Postprocessors Today," *Frontiers in Manufacturing Technology*, vol. II, Institute of Science and Technology, The University of Michigan, 1967, pp. 24–41.
4. ———. *System/360 APT Numerical Control Processor, Version 4—Part Programming Manual*, IBM Corporation, publication GH20-0309, February 1972.
5. ———. *N/C Handbook*, L. J. Thomas, Ed., Bendix Corporation, Industrial Controls Division, 1971, pp. 127–74.

Problems

1. A five-axis machine tool is positioned so that the tip of the cutting tool is at $(0, 0, 0)$ of a fixed reference frame. The angular orientation is $A = 30°$; $B = 45°$. If the tool is 75 mm long, sketch the tool's projection in each of the three orthogonal planes. Label all axes and angles.

2. A two-axis contouring machine is to be used to drill holes at $(0.1, 0.1)$, $(0.2, 0.075)$, and $(0.2, 0.15)$. The feedrate between points is specified to be 1.0 mpm, and each drilling operation requires 4 sec. The tool begins and ends its movement at $(0, 0, 0)$. What percent of the total time required to complete the entire operation is spent in traveling from point to point?

3. Substitute a two-axis point-to-point machine for the operation described in Problem 2. What percent of the total time is spent in traveling from point to point using the positioning machine?

4. A square bolt hole pattern must be drilled. If the tool traverse rate is 5 mpm, compare the time required to machine the pattern for a contouring and a positioning machine. The tool begins and ends at $(0, 0)$, and the lower left hand corner of the pattern is at $(0.4, 0.2)$. What if the corner were at $(0.2, 0.2)$?

5. A pocket 0.25 m long and 0.1 m wide must be milled using a positioning machine. Draw the shortest tool path for a 25-mm diameter cutter with 6-mm overlap. List

the coordinates for each line segment in sequence. Use $(0, 0)$ as the coordinate value at the lower left hand corner of the pocket. The tool begins movement from $(-0.5, 0.25)$.

6. A cutting tool must approximate a 0.25 m radius circular arc through an angle of 60°. If tolerance must be 0.001 m, what is the length of each straight line segment needed to approximate the curve? The tool path must begin and end at the extremities of the arc.

7. Develop a relationship similar to equation (7.4), if the chords used to approximate an arc are tangent to the arc (i.e., an *outside* tolerance is required).

8. The curve $y = 3x^2 + 4x - 8$ is to be machined using a 16-mm diameter cutter. When the cutter is tangent to the curve at point $(1, 2)$, what is the coordinate of the cutter center if the curve is to the left of the cutting tool?

9. A contour defined in Table P.1 must be machined into a workpiece using a ball end-mill. If cutters are available in 2-mm increments, what is the largest tool that can be safely used to cut the contour?

Table P.1.

x (mm)	y (mm)
25	150
64	125
89	100
115	25
125	20
150	25
175	50
230	110
280	145

10. Cutter offset is an important consideration when machining contours. If normals are properly calculated, but the cutter specified is too large, what will happen to the tool path? Illustrate your findings with a sketch.

11. An arc of 25-mm radius with a center point at -12 mm, 25 mm begins at point $(-12, 0)$ and moves counterclockwise for 150°. Specify the incremental values necessary for circular interpolation of a cutter path along this arc.

12. The acute angle illustrated in Figure P.7.1 is to be machined as shown. Develop an expression for the uncut material area as a function of the cutter diameter, D, and included angle α.

13. Develop the cutter location coordinates necessary to machine the component illustrated in Figure P.7.2. Cutter diameter is 2.0 units. Circular interpolation is available, and all other curves may be approximated by line segments.

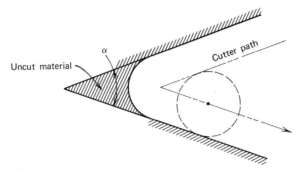

Figure P.7.1   Problem 7.12.

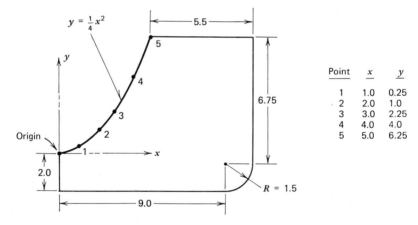

Figure P.7.2   Problem 7.13.

14. Develop the actual NC blocks for the CL data calculated in Problem 13. Use *Magic-3* coding for feedrate and spindle speed. Feedrate is 2 mpm; spindle speed is 300 rpm CCW.

15. A four-axis machine tool uses a rotary table to implement B-axis rotation. If calculations for x, y, z, β are made in a fixed reference frame, a transformation of coordinates must be performed in order to have the proper x and z coordinates supplied to the MCU. Referring to Figure P.7.3, develop the transformation formulae necessary to convert to the machine tool coordinate system from a reference frame that conceptually revolves with the rotary table.

16. Write the SPPL geometry statements necessary to fully define the component illustrated in Figure P.7.4. Attempt to keep the total number of statements to a minimum.

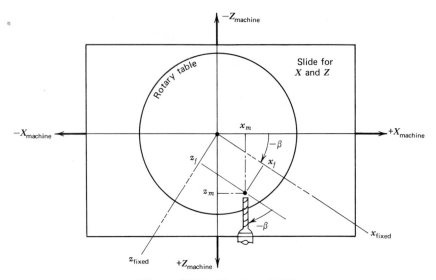

Figure P.7.3   Problem 7.15.

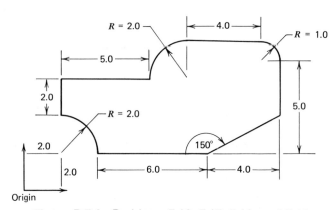

Figure P.7.4   Problems 7.16, 7.17, 7.18, and 7.19.

17. Using the geometry statements from Problem 16, write SPPL motion statements to machine the component illustrated. Generate any additional check lines which may be necessary. Start at the origin and proceed clockwise around the workpiece.

18. For the component shown in Figure P.7.4, how many circular interpolation blocks will be generated by the postprocessor? Develop the incremental coordinates for each block.

19. Assuming that each circular arc in Figure P.7.4 is replaced by a line connecting the points of intersection or tangency, could a point-to-point machine tool be used to

cut the component? If not, what additional design change can be made so that the
component could be machined on a positioning machine?

20. Evaluate the following section of an SPPL program. Draw the geometry described.
    Correct errors that exist in the tool path, and show the proper path on the drawing.

    $\vdots$

    PT1 = POINT/0.0, 0.0

    PT2 = POINT/3.0, 1.0

    PT3 = POINT/4.0, 3.0

    PT4 = POINT/0.0, 3.0

    C1 = CIRCLE/PT2, 1.0

    L1 = LINE/PT1, RIGHT, TANTO, C1

    L2 = LINE/PT3, LEFT, TANTO, C1

    L3 = LINE/PT3, PT4

    L4 = LINE/PT4, PT1

    GFWD/L1, TANTO, C1

    GFWD/C1, TANTO, L2

    GFWD/L2, PAST, L3

    GRIGHT/L3, PAST, L4

    GLEFT/L4, TO, L1

    $\vdots$

21. If the NC programmer did not detect the errors indicated in Problem 20, at what
    point would the computer recognize the error? Explain your answer.

22. The arithmetic element of the SPPL processor is to be expanded to handle a circle
    tangent to two other circles. Evaluate this proposed capability by (1) determining
    how many possible solutions exist for the general case, (2) suggesting a format
    (including new minor words) that could be used to qualify the required circle, and (3)
    illustrating with an example.

23. Develop the analogy for equations 7.10, 7.11, and 7.12 for an acceleration curve of
    the form shown in Figure P.7.5.

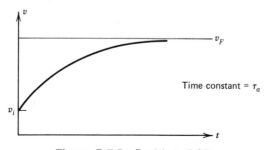

Figure P.7.5 Problem 7.23.

24. A cutting tool is traveling toward a workpiece surface at a feedrate of 3 mpm. When it is 0.2 m from the surface deceleration begins. For a time constant of 0.5 sec, can the tool be brought to rest before it strikes the surface? If not, what is the velocity at impact?

25. A machine control unit manufacturer claims that his exponential deceleration feature ($\tau = 0.675$ sec) enables a tool traveling at 10 mpm to be slowed to 1 mpm in less than 3 sec and in a distance of less than 0.4 m. Are his claims valid?

26. A machine tool is traveling at a constant feedrate of 1.1 mpm. It must decelerate to a cornering feedrate of 0.4 mpm in a distance of 0.1 m in the shortest possible time. What value of output feedrate will achieve this condition?

27. A machine tool is temporarily at rest when a DRILL cycle command, **g81**, is encountered by the MCU. Outline the commands that must be internally generated to perform the cycle operation.

28. Let us assume that a new SPPL processor supports the MACRO, arithmetic statement, and looping logic available in APT. Use the elements to define the function contour $y = \frac{1}{4}x^2$ illustrated in Figure P.7.2.

29. Write an SPPL part program to generate the NC information necessary to machine the component illustrated in Figure P.7.2.

30. Write a generalized MACRO in SPPL that can be used to define the geometry and motion for Figure P.7.6. Note that all coordinates and the circle diameter are variable.

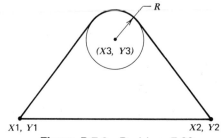

$X1, Y1$ $X2, Y2$

Figure P.7.6 Problem 7.30.

31. Given the following SPPL geometric definition

$$PT1 = POINT/3.0, 6.0, 0.0$$
$$PT2 = POINT/5.0, 2.0, 0.0$$
$$PT3 = POINT/9.0, 4.0, 0.0$$
$$L1 = LINE/PT1, PT2$$
$$L2 = LINE/PT2, PT3$$
$$GFWD/L1, TO, L2$$
$$GLEFT/L2, \ldots$$

illustrate the drive, check, and part surface with the tool axis parallel to the z axis.

Chapter Eight
Mathematics for Numerical Control

Numerical control exhibits a broad range of mathematical applications that include differential equations, integral and operational calculus, state-variable techniques, matrix methods, and many other approaches (References 1 and 2). In this chapter, we are concerned with the types of mathematical operations used to define data input to the MCU.

In the previous chapter examples of simple mathematical operations were presented. To generate the appropriate cutter offset points, the manual part program required only trigonometric relationships. However, as contouring or positioning requirements become more complex, higher forms of analysis are necessary. For example, spatial (3-D) contouring of even relatively simple* shapes requires complicated relationships derived using analytic geometry. This results in the need for computerized analysis methods.

Two important branches of analysis are encountered in the generation of NC part programs. The first, *analytic geometry*, allows the part programmer to write, in APT

$$L1 = LINE/POINT1, LEFT, TANTO, CIRCL3 \qquad (8.1)$$

Expression (8.1) indicates that a line, L1, through a given point and tangent to a given circle is to be defined. Although statements such as this may require relationships that are somewhat complicated, their derivations and subsequent use are straightforward. The second branch of analysis arises when contours like the curve in Figure 8.1 must be machined. The contour is defined by a set of points that cannot *exactly* be defined by an analytical expression. For Figure 8.1, no single function, $y = f(x)$, can be found that precisely describes each point p_i. Therefore, the methods of analytic geometry must be modified before such curves can be analyzed. A third form of analysis must be used when surfaces that cannot be precisely defined by some $f(x, y, z)$ are encountered.

---

* A simple shape is defined as one made up of planar surfaces and well defined solids such as cylinders, cones, and spheres. Even the simple surface can be quite difficult to model.

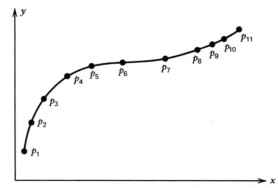

Figure 8.1   A two-dimensional contour defined by 11 discrete points.

There is no universally accepted generic name for analysis methods that consider curves and surfaces described only by discrete points. For purposes of clarity, we refer to them as *contour fitting procedures*.

8.1 Analytic Geometry in NC

Analytic geometry is the mechanism that enables a part programmer to specify cutter position. If the part program is developed by hand, each geometric relationship must be derived and then applied to the data. Computer-aided techniques eliminate the need for extensive derivations. Regardless of the programming technique, analytic geometry provides the means for cutting vector specification.

8.1.1 Line-Arc Evaluation

Consider the simple machining operation illustrated in Figure 8.2. The cutter is to move along the line l_1, starting at point P_1 and terminating at a point tangent to circle C_1. Motion is then to proceed in a counterclockwise direction tangent to the circle and terminate at line l_2.

The cutter motion illustrated in Figure 8.2 can be described only after the tangency point, P_2, to the circle is defined and the terminal point, P_n, tangent to both the circle and the line l_2 is determined. Assuming that circular interpolation is not available, an expression for intermediate points on the arc P_2P_n must be derived. The slope, m, and y-intercept of each line are known (or can be calculated based on related information), and the circle center (X_c, Y_c) and radius, R, are given. The tool radius is r. Because we must calculate points

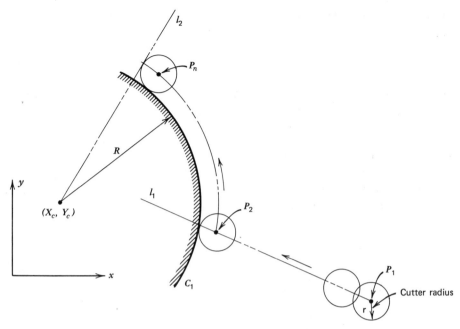

Figure 8.2 Cutter motion to and along a circular arc.

along the arc that may involve mathematical inaccuracy, a tolerance value, t, must also be specified (see Section 7.3).

To calculate the tangency point, P_2, we define a new circle of radius

$$R' = R + r$$

which describes the locus of all circles of radius r (the cutter radius) tangent to C_1. Now, the circle C_1 is described by

$$(x - X_c)^2 + (y - Y_c)^2 = R'^2 \qquad (8.2)$$

and for line l_1:

$$y = m_1 x + b_1 \qquad (8.3)$$

Substituting equation (8.3) into (8.2) yields

$$x^2 + 2xX_c + 2X_c^2 + [m_1^2 x^2 + 2m_1 x(b_1 - Y_c) + (b_1 - Y_c)^2] = R'^2$$

which can be rearranged to yield

$$x^2 + 2\frac{(md - X_c)}{(1 + m_1^2)}x + \frac{2X_c^2 + d^2 - R'^2}{(1 - m_1^2)} = 0 \qquad (8.4)$$

where $d = b_1 - Y_c$. Since equation (8.4) is in standard quadratic form it follows that

$$x_2 = \frac{2X_c - 2m_1 d}{2(1+m_1^2)} + \sqrt{\frac{2X_c^2 + d^2 - R'^2}{4(1+m_1^2)}} \qquad (8.5a)$$

The nature of the geometry dictates that the larger x value be chosen; hence, the positive radical term appears. From equation (8.3)

$$y_2 = m_1 x_2 + b_1 \qquad (8.5b)$$

where x_2 is defined as in equation (8.5a).

To calculate the terminal point, P_n, we note that the cutter circle lies tangent to both C_1 and l_2. If we define a new line, l_2', parallel to line l_2, as shown in Figure 8.3, point P_n can be defined as the intersection of C_1' and l_2'. Referring to the figure, angle α is defined such that

$$m_2 = \tan \alpha$$

To calculate the equation for the new line l_2' we must determine the new y intercept, b_2'. From the figure, it can be seen that

$$\Delta y = r/\cos \alpha = r \sec \alpha$$

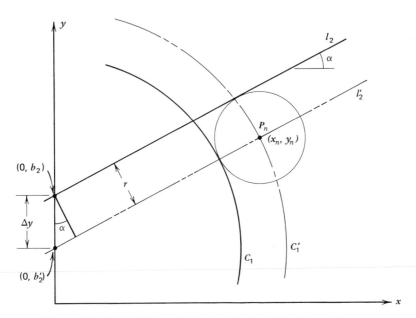

Figure 8.3 Calculation of the terminal point, P_n, on the arc.

Now, by definition

$$\sec \alpha = \sqrt{\tan^2 \alpha - 1} = \sqrt{m^2 - 1}$$

Therefore

$$b_2' = b_2 - \Delta y = b_2 - r\sqrt{m^2 - 1}$$

The equation for l_2' can now be written

$$y = m_2 x + b_2' \qquad (8.6)$$

and point P_n can be calculated using the approach outlined in equations (8.2) to (8.5b), substituting equation (8.6) for (8.3) and proceeding as shown. Hence, the end points of the arc have been defined.

Before continuing with a discussion of the intermediate points on the path, a number of important comments should be made. The method presented for the calculation of points P_2 and P_n is not the only approach to the problem. Computerized analysis methods generally use the *canonical* forms for geometric entity definitions. For example, the canonical form for a line is

$$ax + by + cz + d = 0$$

This form is most general and alleviates the problems of infinite slope. It should also be noted that in carrying out the derivations, we performed operations based on visualization; for example, the line l_2 was offset in the negative y direction and the positive radical was chosen in equation (8.5a) because we *saw* that the geometry required these steps. In computer-aided NC programming, keywords must be used to allow the processor to make similar logical decisions.

8.1.2 Defining an Arc

To define intermediate points on the arc P_2P_n, an angle, ϕ, must be determined such that the tolerance value is maintained. Three methods may be used to define tolerance as illustrated in Figure 8.4. In Chapter Seven, we defined ϕ for a chordal tolerance to be

$$\phi = 2 \cos^{-1}(1 - t/R) \qquad (7.3)$$

Similarly, it can be shown that the value of ϕ for tangential tolerance is

$$\phi = 2 \cos^{-1}\left(\frac{R}{t + R}\right)$$

and for secantial tolerance

$$\phi = 2 \cos^{-1}[(1 + t/R)/(1 - t/R)]$$

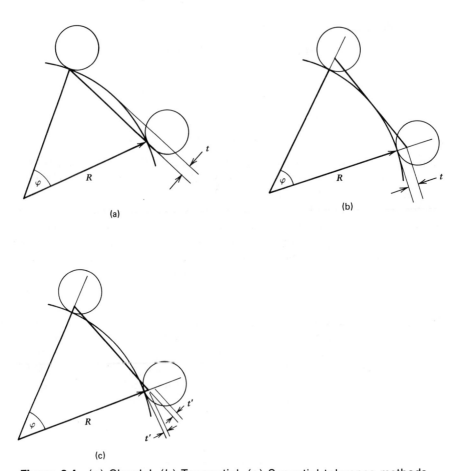

Figure 8.4 (a) Chordal. (b) Tangential. (c) Secantial tolerance methods.

Given ϕ we can now define any number of intermediate points based on the geometries illustrated in Figure 8.4. To calculate points on the arc P_2P_n it can be shown that

$$x_i = X_c + \left(R + r \sec \frac{\phi}{2}\right) \sin (\alpha_2 + k\phi) \qquad (8.6a)$$

$$y_i = Y_c + \left(R + r \sec \frac{\phi}{2}\right) \cos (\alpha_2 + k\phi) \qquad (8.6b)$$

where $\alpha_2 = \tan^{-1} m_2$; $i = 3, 4, \ldots, n - 1$; and $k = 1, 2, \ldots, m$. Hence, the intermediate points, P_i, have been derived. The tangential and secantial methods of approximation result in somewhat different formulae. It is left to the reader to derive these.

It is important to note that the previous expressions for the angle ϕ are valid only for a cutter traveling on the convex side of a circular arc. The chordal tolerance for a cutter moving along the inside (concave side) of a circular arc is illustrated in Figure 8.5. Defining t as the tolerance for a cutter of radius r, and t_0 as the tolerance for a zero-radius cutter, we can write,

$$t_0 = R\left(1 - \cos\frac{\phi}{2}\right) \tag{8.7a}$$

$$t_0 - t = r\left(1 - \cos\frac{\phi}{2}\right) \tag{8.7b}$$

Solving equations (8.7) simultaneously,

$$\cos\frac{\phi}{2} = \frac{\Delta r - t}{\Delta r} = 1 - \frac{t}{\Delta r}$$

where $\Delta r = R - r$, the difference in radii. Hence,

$$\phi = 2\cos^{-1}\left(1 - \frac{t}{\Delta r}\right) \tag{8.8}$$

The difference between equation (8.8) and (7.3) is that the angle specification for internal tolerances is a function of t, R, *and* r.

The examples discussed above illustrate simple situations. If the curve under consideration were parabolic instead of circular, the development would be

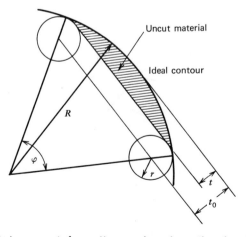

Figure 8.5 Inner tolerance, t, for cutter moving along the chord of a concave arc.

more difficult. For higher order curves, the analysis generally resorts to iterative procedures for the solution of intersections and proper tolerances for contouring.

8.2 Computer-Aided Analysis

When a cutter must be positioned with regard to surfaces, as opposed to simple two-dimensional contours, the use of equations to calculate curve intersections and cutter offsets is insufficient. Computer-aided analysis techniques make use of a general iterative program which produces the *cut vector* based on specific surface and tool data. This program, called the *arithmetic element* (ARELEM) in the APT programming language, has already been discussed in general terms. In this section we consider the basic analysis method that enables the tool motion to be determined.

The objective of the ARELEM iterative procedure is to move the tool along a path that (1) keeps the tool *tangent* to both the drive surface and the part surface, (2) is defined in increments short enough so that tolerance is maintained, and (3) obeys constraints implied by various check surfaces. It should

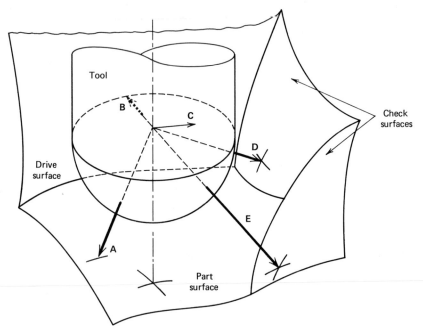

Figure 8.6   Formulation of *cut vector* and *pull vectors* in ARELEM (Reference 3).

be recalled that the drive surface guides the tool along the workpiece (part) surface. Check surfaces are used to terminate or modify cutter motion. The arithmetic element calculates a linear cutter motion that best approximates the surface.

The ARELEM iterative analysis depends on *minimum distance vectors*, or *pull vectors* (Reference 3), illustrated in Figure 8.6. In the figure, **A** and **B** are the minimum distance vectors to the drive and part surfaces, respectively. Vector **C** is the component of motion along the cutter path. The other vectors are pull vectors corresponding to various check surfaces. At any instant in time, each surface exerts a *pull* on the cutter path.

The pull vectors are calculated using an iterative procedure, illustrated in Figure 8.7. Since each pull vector represents a minimum distance between the tool envelope and the surface, points on the tool and surface are chosen so that their respective normals become approximately colinear. When the normals lie within some specified angular tolerance, they are considered properly

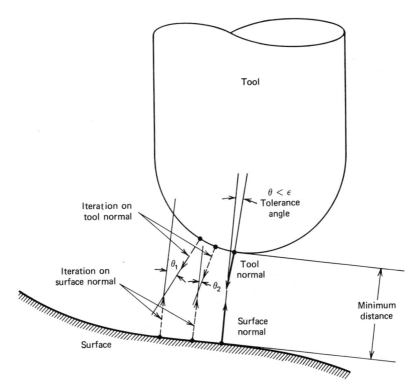

Figure 8.7   Iteration scheme for calculation of pull vector.

aligned. The alignment direction defines the minimum distance pull vector. Once these vectors have been obtained, **C** may be determined.

To determine an initial guess for the proper vector direction along the cutter path, the cross product of **A** and **B** is taken, resulting in a vector direction illustrated by line l in Figure 8.8. Once this direction has been established, **C** can be obtained by projecting each of the pull vectors to line l, and choosing the *shortest* component (in this case, vector **D**, projected). The shortest vector is selected because any longer motion would cause the cut to exceed tolerance on any surface having a pull vector which results in a shorter projection onto l.

The desired cutting tool motion is obtained by combining each pull vector component (i.e., **A**, **B**, and **C**) into a resultant motion. The magnitude of each of the components is modified to conform to tangency requirements. A simplified version of the iteration is illustrated in Figure 8.9. At point P_1 the vector C_1 is

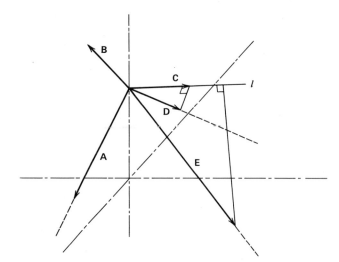

Figure 8.8   Determination of the vector, **C**.

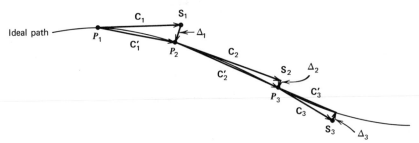

Figure 8.9   A simplified technique for path determination.

calculated using the methods oulined above. Pull vectors are calculated at S_1 and a resultant correction vector, Δ_1, is calculated. Now

$$\mathbf{C}_1' = \mathbf{C}_1 + \Delta_1$$

where $\mathbf{C}_1'$ is the actual cutter path. The process continues at each point. It should be noted that the length of the vector $\mathbf{C}_1$ is chosen so that the tolerance value is never exceeded.

In practice, the calculations within an arithmetic element can become quite complicated. For a complete discussion of the computer programs used for NC part program analysis, the reader is referred to specific program documentation (Reference 4).

8.3 Contour Fitting Procedures

Although many NC machined contours, such as the line, arc, circle, ellipse, and parabola, conform to analytically defined mathematical equations, a growing number of applications require *free-form* shapes that do not conform to any single functional relationship. For such contours to be machined using numerical control, a method for the mathematical description of curves like that in Figure 8.1 must be available.

A draftsman draws a smooth curve through a set of points using a flexible plastic, metal, or wooden device called a *spline*. The spline, illustrated in Figure 8.10, is bent through predefined points (*knots*) and anchored by weights. These weights, sometimes called *ducks*, are strategically placed so that the desired curve is outlined. This concept of a physical spline has led to the development of *mathematical splines*.

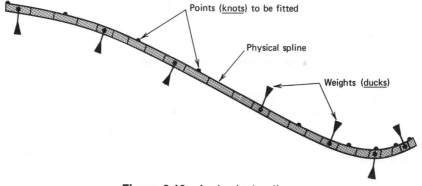

Points (<u>knots</u>) to be fitted

Physical spline

Weights (<u>ducks</u>)

Figure 8.10 A physical spline.

8.3.1 The Mathematical Spline

The physical spline assumes a shape such that its internal strain energy is minimized. We can express this condition mathematically in the following manner. Strain energy, u_e, may be defined as

$$u_e = K \int_0^L \rho^2 \, ds \qquad (8.9)$$

where K is the structural stiffness of the spline (a constant), ρ is the curvature, and ds is an incremental arc length for $0 \leq s \leq L$. In the Cartesian reference frame it can be shown that (Reference 5)

$$u_e = K \int \frac{(d^2y/dx^2)^2}{(1 + dy/dx)^{5/2}} \, dx \qquad (8.10)$$

Equation (8.10) can be simplified if we assume that $dy/dx \ll 1$. Then

$$u_e = K \int \left(\frac{d^2y}{dx^2}\right)^2 dx \qquad (8.11)$$

Minimization of equation (8.11), with a constraint requiring that the resultant curve pass through a set of predefined points, leads to series of cubic curves connected at the points. The cubic curves must be continuous in position, slope, and curvature at each knot. Hence, the curve is divided into n intervals, each described by a different cubic function.

The mathematical spline produces results that compare favorably with the physical spline. However, if the curve to be *spline fit* exhibits large slope values, then the assumption, $dy/dx \ll 1$, no longer holds and problems can result. This situation may sometimes be remedied by rotating the curve so that a large slope becomes small with respect to another set of axes.

To eliminate slope constraints, a more general spline formulation is used. The *parametric spline* defines the curve in terms of an independent parameter that describes the contour in vector notation.

8.3.2 The Parametric Spline

The parametric spline has a number of important characteristics that provide distinct advantages in relation to the mathematical model of a physical spline:

1. A parametric spline is essentially axis independent.
2. Multivalued (i.e., looped) curves can be defined, and no slope constraints exist.
3. Fully three-dimensional space curves can be analyzed.

It should be noted that unlike the mathematical spline, the parametric spline does not satisfy minimum energy. However, the curves are smooth and continuous.

Many varieties of parametric splines have been proposed in the literature (e.g., References 6–8). As an illustrative example, we consider the Ferguson curve (Reference 6), which has been the foundation for many subsequent methods.

Referring to Figure 8.11a, let **A** and **B** be two vectors with end points a and b, and vectors $\mathbf{T}_A$ and $\mathbf{T}_B$ tangent at a and b, respectively. We can define a space curve through a and b, tangent to $\mathbf{T}_A$ and $\mathbf{T}_B$, that has the following form:

$$\mathbf{p}(u) \equiv \sum_{i=0}^{3} \mathbf{R}_i u^i \tag{8.12}$$

where u is a nondimensional *parameter* taking the values $0 \le u \le 1$. The function $\mathbf{p}(u)$ has the following boundary conditions:

$$\mathbf{p}(0) = \mathbf{A}; \quad \mathbf{p}(1) = \mathbf{B}; \quad \left.\frac{d\mathbf{p}}{du}\right|_{u=0} = \mathbf{T}_A; \quad \left.\frac{d\mathbf{p}}{du}\right|_{u=1} = \mathbf{T}_B$$

By substituting these values into equation (8.12), it can be shown that

$$\mathbf{R}_0 = \mathbf{A}$$

$$\mathbf{R}_1 = \mathbf{T}_A$$

$$\mathbf{R}_2 = 3(\mathbf{B} - \mathbf{A}) - 2\mathbf{T}_A - \mathbf{T}_B$$

$$\mathbf{R}_3 = 2(\mathbf{A} - \mathbf{B}) + \mathbf{T}_A + \mathbf{T}_B$$

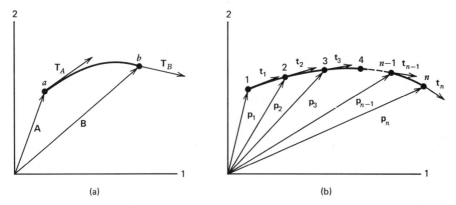

Figure 8.11 Two-dimensional representation of a parametric spline. (a) Single interval. (b) Generalized curve.

Hence, equation (8.12) can be written

$$\mathbf{p}(u) = u^3[2(\mathbf{A} - \mathbf{B}) + \mathbf{T}_A + \mathbf{T}_B] + u^2[3(\mathbf{B} - \mathbf{A}) - 2\mathbf{T}_A - \mathbf{T}_B] + u\mathbf{T}_A + \mathbf{A}$$
(8.13)

which describes the curve between $\mathbf{A}$ and $\mathbf{B}$ in terms of the parameter u. Because equation (8.13) defines a general interval between two points, it can easily be adapted to a set of n points with $n - 1$ intervals.

Let $\mathbf{p}_j$ be n vectors in space, and $\mathbf{t}_j$ be the corresponding tangent vectors, as illustrated in Figure 8.11b. For the jth interval equation

$$\mathbf{p}_j(u) = u^3[2(\mathbf{p}_j - \mathbf{p}_{j+1}) + \mathbf{t}_j + \mathbf{t}_{j+1}] + u^2[3(\mathbf{p}_{j+1} - \mathbf{p}_j) - 2\mathbf{t}_j - \mathbf{t}_{j+1}] + u\mathbf{t}_j + \mathbf{p}_j \quad (8.14)$$

For the curve to be smooth and continuous, the following conditions must be satisfied at the jth interval:

$$\mathbf{p}_j(1) = \mathbf{p}_{j+1}(0)$$
$$\mathbf{p}'_j(1) = \mathbf{p}'_{j+1}(0)$$
$$\mathbf{p}''_j(1) = \mathbf{p}''_{j+1}(0)$$
(8.15)

where the single and double primes indicate first and second derivatives with respect to u. Rewriting equation (8.14)

$$\mathbf{p}_j(u) = \mathbf{R}_{3j}u^3 + \mathbf{R}_{2j}u^2 + \mathbf{R}_{1j}u + \mathbf{R}_{0j}$$

it follows that

$$\mathbf{p}'_j(u) = 3\mathbf{R}_{3j}u^2 + 2\mathbf{R}_{2j}u + \mathbf{R}_{1j}$$

and

$$\mathbf{p}''_j(u) = 6\mathbf{R}_{3j}u + 2\mathbf{R}_{2j}$$

Substituting the above expressions into equations (8.15) and evaluating at $u = 1$, the following equation results (after simplification):

$$\mathbf{t}_j + 4\mathbf{t}_{j+1} + \mathbf{t}_{j+2} = 3(\mathbf{p}_{j+2} - \mathbf{p}_j) \quad (8.16)$$

Equation (8.16) provides a generalized relationship between point and tangent vectors on the curve. Examination of conditions (8.15) indicates that only the second through $(n - 1)$th point are considered. At the end points, p_1 and p_n, we can specify

$$\mathbf{p}''_1(0) = 0; \quad \mathbf{p}''_{n-1}(1) = 0$$

which indicates that zero curvature is assumed at the end points.* This condition yields two more equations:

$$2\mathbf{t}_1 + \mathbf{t}_2 = 3(\mathbf{p}_2 - \mathbf{p}_1)$$
$$\mathbf{t}_{n-1} + 2\mathbf{t}_n = 3(\mathbf{p}_n - \mathbf{p}_{n-1})$$
(8.17)

---

* Zero curvature is assumed for simplicity only. Actually, any values may be used.

Equations (8.16) and (8.17) comprise n equations in n unknowns. All values $\mathbf{p}_j$ are known, so that we have a solvable system of simultaneous equations from which the unknown tangent vectors are obtained. In general

$$2\mathbf{t}_1 + \mathbf{t}_2 + 0 + 0 + 0 + 0 + \cdots = 3(\mathbf{p}_2 - \mathbf{p}_1)$$
$$\mathbf{t}_1 + 4\mathbf{t}_2 + \mathbf{t}_3 + 0 + 0 + 0 + \cdots = 3(\mathbf{p}_3 - \mathbf{p}_1)$$
$$0 + \mathbf{t}_2 + 4\mathbf{t}_3 + \mathbf{t}_4 + 0 + 0 + \cdots = 3(\mathbf{p}_4 - \mathbf{p}_2)$$
$$0 + 0 + \mathbf{t}_3 + 4\mathbf{t}_4 + \mathbf{t}_5 + 0 + \cdots = 3(\mathbf{p}_5 - \mathbf{p}_3)$$
$$\vdots \qquad\qquad\qquad \vdots \qquad (8.18)$$
$$0 + 0 + 0 + \cdots + \mathbf{t}_{n-2} + 4\mathbf{t}_{n-1} + \mathbf{t}_n = 3(\mathbf{p}_n - \mathbf{p}_{n-2})$$
$$0 + 0 + 0 + \cdots + 0 + \mathbf{t}_{n-1} + 2\mathbf{t}_n = 3(\mathbf{p}_n - \mathbf{p}_{n-1})$$

Since the above system of equations is nearly triangular, a solution can be easily accomplished.

Throughout the development leading to equations (8.18), we have represented both points and tangents in vector form. Referring to Figure 8.12,* it is evident that

$$\mathbf{p}_j = x_j \mathbf{u}_1 + y_j \mathbf{u}_2 + z_j \mathbf{u}_3 \qquad (8.19a)$$

and

$$\mathbf{t}_j = t_{xj} \mathbf{u}_1 + t_{yj} \mathbf{u}_2 + t_{zj} \mathbf{u}_3 \qquad (8.19b)$$

---

* For simplicity in visualization, the z component is ignored in the figure.

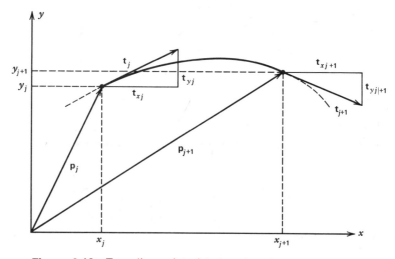

Figure 8.12 Two-dimensional tangent vector components.

where $\mathbf{u}_1$, $\mathbf{u}_2$, and $\mathbf{u}_3$ are unit vectors in the x, y, and z directions. If the right hand side of equations (8.19) is substituted into equations (8.18), three sets of n equations will result. That is, the tangent components in each principle direction are solved independently. Once the tangent components are known, the appropriate curve can be generated using equation (8.14) for each interval.

8.3.3 An Example of the Parametric Spline

Given the set of points $(0.1, 0.1)$, $(0.7, 0.45)$, $(1.2, 0.3)$ on the x-y plane, the Ferguson parametric spline may be used to determine the tangent vectors at each knot of the spline. For this simple example, equations (8.18) reduce to

$$2\mathbf{t}_1 + \mathbf{t}_2 = 3(\mathbf{p}_2 - \mathbf{p}_1)$$
$$\mathbf{t}_1 + 4\mathbf{t}_2 + \mathbf{t}_3 = 3(\mathbf{p}_3 - \mathbf{p}_1)$$
$$\mathbf{t}_2 + 2\mathbf{t}_3 = 3(\mathbf{p}_3 - \mathbf{p}_2)$$

In matrix form we can write

$$\begin{bmatrix} 2 & 1 & 0 \\ 1 & 4 & 1 \\ 0 & 1 & 0 \end{bmatrix} \begin{bmatrix} \mathbf{t}_1 \\ \mathbf{t}_2 \\ \mathbf{t}_3 \end{bmatrix} = 3 \begin{bmatrix} \mathbf{p}_2 - \mathbf{p}_1 \\ \mathbf{p}_3 - \mathbf{p}_1 \\ \mathbf{p}_3 - \mathbf{p}_2 \end{bmatrix} \qquad \text{(i)}$$

Equation (i) can be rewritten

$$[A] \cdot [T] = 3[\delta P]$$

or   (ii)

$$[T] = [A]^{-1}[\delta P]$$

Now, the inverse of the coefficient matrix, $\mathbf{A}^{-1}$, can be determined using standard matrix techniques (e.g., see References 9 and 10). Hence

$$[A]^{-1} = \frac{1}{12} \begin{bmatrix} 7 & -1 & 1 \\ -2 & 4 & -2 \\ 1 & -2 & 7 \end{bmatrix} \qquad \text{(iii)}$$

In order to solve for the tangent vectors at the knots of the spline, equation (iii) must be expressed in terms of the vector components t_x and t_y. Using equations (i) and (iii)

$$\begin{bmatrix} t_{x1} \\ t_{x2} \\ t_{x3} \end{bmatrix} = \frac{1}{12} \begin{bmatrix} 7 & -1 & 1 \\ -2 & 4 & -2 \\ 1 & -2 & 7 \end{bmatrix} 3 \begin{bmatrix} .7 - 0.1 \\ 1.2 - 0.1 \\ 1.2 - 0.7 \end{bmatrix}$$

which can be evaluated to yield

$$t_{x1} = 0.9; \quad t_{x2} = 0.55; \quad t_{x3} = 0.475$$

Likewise, the y components can be expressed as

$$\begin{bmatrix} t_{y1} \\ t_{y2} \\ t_{y3} \end{bmatrix} = \frac{1}{12} \begin{bmatrix} 7 & -1 & 1 \\ -2 & 4 & -2 \\ 1 & -2 & 7 \end{bmatrix} 3 \begin{bmatrix} 0.45 - 0.1 \\ 0.3 & -0.1 \\ 0.3 & -0.45 \end{bmatrix}$$

which can be evaluated to yield

$$t_{y1} = 0.525; \quad t_{y2} = 0.1; \quad t_{y3} = -0.275$$

Recalling equation (8.19b),

$$t_j = t_{xj}\mathbf{u}_1 + t_{yj}\mathbf{u}_2$$

Therefore, the tangent vectors at the spline knots are:

$$\mathbf{t}_1 = 0.9 \quad \mathbf{u}_1 + 0.525\mathbf{u}_2$$
$$\mathbf{t}_2 = 0.55 \quad \mathbf{u}_1 + 0.1 \quad \mathbf{u}_2$$
$$\mathbf{t}_3 = 0.475\mathbf{u}_1 - 0.275\mathbf{u}_2$$

It is important to note that matrix techniques have been used to illustrate the procedure required for actual problems. The matrix $\mathbf{A}^{-1}$ can be found using a banded matrix solution which greatly reduces computation time.

8.3.4 Interpolation and Normals

A parametric spline, described by the point and tangent vectors discussed above, is a contour that must be followed by a numerically controlled tool. If we consider an NC contouring device with linear interpolation, the cutter path becomes a curve parallel to the spline. This curve consists of small line segments that divide each interval into an appropriate number of subintervals.

The vector function $\mathbf{p}_j(u)$ defines a parametric spline interval between p_j and p_{j+1}. A point on the curve may be calculated by specifying the appropriate u value such that $0 \leq u \leq 1$. The parameter u can be viewed as a nondimensionalized measure of arc length, as illustrated in Figure 8.13. Hence, any number of intermediate points (i.e., points between p_j and p_{j+1}) may be defined simply by substituting the appropriate value of u into equation (8.14).

For example, let us assume that a point midway along the curve segment (Figure 8.13) is desired. Since points p_j are given, $\mathbf{t}_j$ can be computed using equations (8.18). To obtain a point midway along the segment, let $u = 0.5$. Then on the jth interval in two dimensions:

$$p_x(0.5) = 0.125[2(x_j - x_{j+1}) + t_{xj} + t_{xj+1}] + 0.25[3(x_{j+1} - x_j) - 2t_{xj} - t_{xj+1}] + 0.5t_{xj} + x_j$$

$$p_y(0.5) = 0.125[2(y_j - y_{j+1}) + t_{yj} + t_{yj+1}] + 0.25[3(y_{j+1} - y_j) - 2t_{yj} - t_{yj+1}] + 0.5t_{yj} + y_j$$

Since values of x_j, x_{j+1}, y_j, y_{j+1}, and all tangent vector components are known, p_x

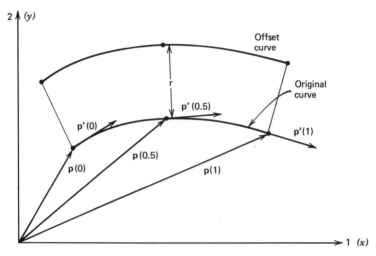

Figure 8.13 Interpolation and normal calculations.

and p_y define a point midway along the arc. In this way we can *interpolate* any number of points on the curve segment.

The offset or parallel curve can also be defined making use of the properties of the function $p(u)$. From equation (8.14),

$$p_j'(u) = \frac{d}{du} p_j(u) = 3u^2[2(\mathbf{p}_j - \mathbf{p}_{j+1}) + \mathbf{t}_j + \mathbf{t}_{j+1}] + 2u[3(\mathbf{p}_{j+1} - \mathbf{p}_j) - 2\mathbf{t}_j - \mathbf{t}_{j+1}] + \mathbf{t}_j$$

which for $0 \leqslant u \leqslant 1$ yields the tangent vector at any point along the curve segment. Consider the two-dimensional components of the function $\mathbf{p}_j'(u)$, $p_{1j}'(u)$ and $p_{2j}'(u)$, where axes 1 and 2 are perpendicular. By definition the direction cosines of a normal to the curve at $u = u_0$ are (dropping the j):

$$l_1(u_0) = -p_2'(u_0)/D$$
$$l_2(u_0) = -p_1'(u_0)/D \tag{8.20}$$

where

$$D = \{[p_1'(u_0)]^2 + [p_2'(u_0)]^2\}^{1/2}$$

Once the direction cosines of the normal have been calculated, the point on the offset curve corresponding to $p_j(u_0)$ may be determined in the following manner:

$$p_1(u_0)_{\text{offset}} = p_1(u_0)_{\text{orig}} \pm l_1(u_0) \cdot r$$
$$p_2(u_0)_{\text{offset}} = p_2(u_0)_{\text{orig}} \pm l_2(u_0) \cdot r \tag{8.21}$$

where r is the offset distance (e.g., a cutting tool radius), and the $\pm$ sign indicates offset *above* or *below* the original curve.

8.3.5 An Example of Interpolation and Offset

For the curve described in Section 8.3.3, a point that has an x coordinate in the range $0.9 \leqslant x \leqslant 1.1$ can be interpolated. Using equations (8.14) and (8.21) we can determine the coordinates of a point that lies on a normal through the interpolated point at a distance $r = 0.2$.

Since the tangent vectors have been calculated in Section 8.3.3, equation (8.14) can be used directly to interpolate the point. Referring to Figure 8.14, it is seen that in the range $0.9 \leqslant x \leqslant 1.1$, $u \approx 0.5$, and $j = 2$. Therefore,

$$\mathbf{p}(u) = u^3[2(\mathbf{p}_2 - \mathbf{p}_3) + \mathbf{t}_2 + \mathbf{t}_3] + u^2[3(\mathbf{p}_3 - \mathbf{p}_2) - \mathbf{t}_2 - \mathbf{t}_3] + u\mathbf{t}_2 + \mathbf{p}_2$$

Putting the above equation into component form and substituting values previously obtained,

$$p_x(0.5) = 0.125[2(0.7 - 1.2) + 0.55 + 0.475] + 0.25[3(1.2 - 0.7) - 2(0.55) - 0.475]$$
$$+ 0.5(0.55) + 0.7$$

$$p_x(0.5) = 0.96$$

Likewise

$$p_y(0.5) = 0.125[2(0.45 - 0.3) + 0.1 - 0.275] + 0.25[3(0.3 - 0.45) - 0.2 + 0.275]$$
$$+ 0.5(0.1) + 0.45)$$

$$p_y(0.5) = 0.42$$

Hence, the interpolated point is (0.96, 0.42).

To determine the normal through the curve at point (0.96, 0.42), the expression $\mathbf{p}'(0.5)$ must be evaluated to obtain the tangent vector components

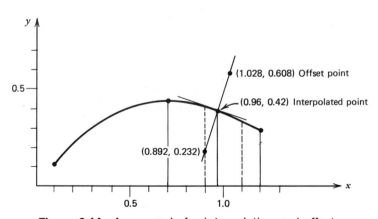

Figure 8.14 An example for interpolation and offset.

at the point. The equations (8.20) and (8.21) must be used. We know that

$$\mathbf{p}_2'(u) = 3u^2[2(\mathbf{p}_2 - \mathbf{p}_1) + \mathbf{t}_2 + \mathbf{t}_3] + 2u[3(\mathbf{p}_3 - \mathbf{p}_2) - 2\mathbf{t}_2 - \mathbf{t}_3] + \mathbf{t}_2$$

Substituting $u = 0.5$ into the above equation and solving for the vector components:

$$p_x'(0.5) = 0.493$$
$$p_y'(0.5) = -0.181$$

With the tangent components known, the direction cosines may be computed using equations (8.20)

$$l_1(0.5) = \frac{0.181}{[(0.493)^2 + (-0.181)^2]^{1/2}} = 0.344$$

$$l_2(0.5) = \frac{0.493}{[(0.493)^2 + (-0.181)^2]^{1/2}} = 0.938$$

The offset point follows from equations (8.21):

$$x_{\text{offset}} = 0.96 \pm 0.0688$$
$$y_{\text{offset}} = 0.42 \pm 0.1875$$

Hence, the two points which are on a line normal to the spline at a distance $r = 0.2$ from the curve are $(0.892, 0.232)$ and $(1.028, 0.608)$.

8.3.6 NC Implementation of Splines

Although the spline mathematics presented in the preceding paragraphs is not conceptually difficult, the application of equations (8.14), (8.18), (8.20), and (8.21) requires a large number of numerical computations for even a small number of input knots. For this reason, splines are generally calculated using computerized methods in an NC part programming language.

The NC part programmer need only specify the original knots on the contour to be machined. All analysis is performed within the arithmetic element. To illustrate the manner in which a spline might be defined, let us add a spline geometry statement to the SPPL, the part programming language developed in Chapter 7.

The part programmer could write

$$spline_1 = \text{SPLINE}/pt_1, pt_2, pt_3 \cdots pt_n$$

where pt_i are predefined points whose coordinates become the component values for $\mathbf{p}_i$ (equation 8.14). Solution of tangents, interpolation, and normal calculation are all performed within the arithmetic element.

8.4 Surfaces

Two types of dimensional surfaces are encountered in numerical control applications. The first, a surface that can be defined in (x, y, z) by a single equation, is discussed in this section. The second, a surface described by a grid of points in space, shall be considered in Section 8.5.

8.4.1 Analytic Surfaces

A surface that is described by a single equation can be termed an analytic surface. Given any two coordinates, the third can be determined; calculation of surface slopes, normals, and curvatures is straightforward; by definition, every point on the surface is exactly defined; that is, interpolation is not necessary. A surface with the above properties can be evaluated using analytic geometry or metric differential geometry (Reference 11). In the following discussion, we present some of the more common surface configurations.

Table 8.1
Quadric Surfaces[a]

Paraboloid	$x^2 + y^2 = 2rz$
Cylinder[b]	$x^2 + y^2 = d^2$
Sphere	$x^2 + y^2 + z^2 = d^2$
Elliptic cone	$\dfrac{x^2}{a^2} + \dfrac{y^2}{b^2} + \dfrac{z^2}{c^2} = 0$
Elliptic cylinder[b]	$\dfrac{x^2}{a^2} + \dfrac{y^2}{b^2} = 1$
Ellipsoid	$\dfrac{x^2}{a^2} + \dfrac{y^2}{b^2} + \dfrac{z^2}{c^2} = 1$
Elliptic paraboloid	$\dfrac{x^2}{a^2} + \dfrac{y^2}{b^2} - 2rz = 0$
Hyperboloid (1 sheet)	$\dfrac{x^2}{a^2} + \dfrac{y^2}{b^2} - \dfrac{z^2}{c^2} = 1$
Hyperboloid (2 sheets)	$\dfrac{x^2}{a^2} - \dfrac{y^2}{b^2} - \dfrac{z^2}{c^2} = 1$
Hyperbolic paraboloid	$\dfrac{x^2}{a^2} - \dfrac{y^2}{b^2} = 2rz$

[a] All surfaces are centered at the origin.
[b] Centerline lies on z-axis.

The most common surface encountered in NC machining applications is the plane. A plane is defined by the following linear equation

$$Ax + By + Cz - D = 0 \qquad (8.22)$$

where A, B, C, and D are constants that describe the surface orientation. For example, a plane perpendicular to the x-axis, at a distance 5 units from the origin, is described by $B = C = 0$, and $D/A = 5$.

The general quadric surface is the locus of points in three dimensions that satisfy the following second degree equation:

$$Ax^2 + By^2 + Cz^2 + 2Fyz + 2Gxz + 2Hxy + 2Px + 2Qy + 2Rz - D = 0$$
$$(8.23)$$

Although the above expression seems quite formidable, surfaces encountered in NC applications are generally degenerate forms of equation (8.23). For example, a sphere is defined when $F = G = H = P = Q = R = 0$, and $A = B = C = 1$. That is

$$x^2 + y^2 + z^2 = D$$

where $D^{1/2}$ is the radius. Table 8.1 lists common degenerate forms of equation (8.23).

8.4.2 NC Implementation of Analytic Surfaces

In NC applications the analytic surface is generally evaluated using computer assisted techniques. Like other geometric entities, an analytic surface can be specified with a geometry statement in the appropriate NC part programming language.

The APT language provides separate keywords for surfaces that are encountered most often. For example,

$$plane := \text{PLANE}/a, b, c, d$$
$$sphere := \text{SPHERE/CENTER}, x, y, z, \text{RADIUS}, r$$
$$cylinder := \text{CYLNDR}/x, y, z, l_1, l_2, l_3, r$$

In the above APT statements, the plane is defined by direct specification of the constants for equation (8.22). The sphere is defined through specification of its center point and radius, and the cylinder by a point on the axis (x, y, z), the direction cosines of the cylinder axis (l_1, l_2, l_3), and its radius, r. Other statement forms may also be used.

A general quadric surface may be defined in APT using a statement of the form

$$quadric = \text{QADRIC}/a, b, c, f, g, h, p, q, r, d$$

where the data values $(a, b, c, f, \ldots, d)$ correspond to the constants in equation (8.23). For example, an elliptic paraboloid (Table 8.1), could be specified.

$$elliptic\ paraboloid = \text{QADRIC}/C_1, C_2, 0, 0, 0, 0, 0, 0, C_3, 0$$

where $C_1 = 1/a^2$, $C_2 = 1/b^2$, and $C_3 = r$.

8.4.3 An Example of the QADRIC Statement

Because equation (8.23) is completely general, it can be used to define two-dimensional curves as well as three-dimensional surfaces. Using the APT geometry statement QADRIC, the contour illustrated in Figure 8.15 can be defined in the following manner.

From the equation of a parabola, the curve illustrated in the figure can be represented by

$$(y - a)^2 = 2p(x - b)$$

or rearranging

$$y^2 - 2ay - 2px + (a^2 + 2pb) = 0$$

Therefore, if we let

$$C_1 = -a,\ C_2 = -p,\ C_3 = a^2 + 2pb$$

the APT statement for a parabola can be written

$$parabola = \text{QADRIC}/0, 1, 0, 0, 0, 0, C_2, C_1, 0, -C_3$$

It should be evident from the above problem that the QADRIC statement can be used to specify any quadratic curve.

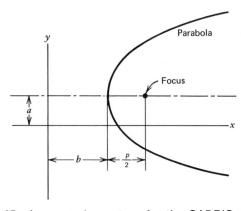

Figure 8.15   An example contour for the QADRIC statement.

8.5 The Sculptured Surface

Unlike analytic surfaces, the sculptured (free-form) surface *cannot* be described by a single mathematical relationship. Methods have been devised for describing such surfaces in a manner that provides the necessary detail for NC applications.

The development of sculptured surface mathematics was due in part to numerical control. In the early 1960s, the design of complex surfaces for aerospace components precluded the use of NC. Although NC contouring machines could follow any surface, efficient means for mathematical description of free-form surfaces were not available. For this reason, a number of independent investigators began the development of sculptured surface techniques.

Two methods (with numerous variations) for the description of sculptured surfaces have been developed. Early investigators (References 12 and 6) used *multivariable curve interpolation* procedures that are roughly analogous to the parametric splining methods discussed earlier. More recent investigations (References 13 and 14) have resulted in *bi-cubic patch* techniques.

8.5.1 Multivariable Curve Interpolation

Multivariable curve interpolation is a natural extension of the parametric splining techniques discussed in Section 8.3. Continuing our discussion of Ferguson's (Reference 6) work, consider an $m \times n$ array of points in three dimensions, as illustrated in Figure 8.16. A smooth surface passing through

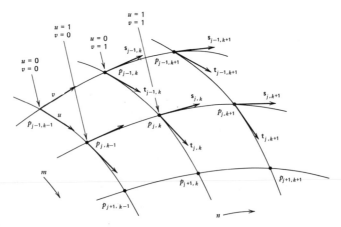

Figure 8.16 Ferguson surface nomenclature (Reference 6).

each point, $p_{j,k}$, and tangent to vectors $\mathbf{t}_{j,k}$ and $\mathbf{s}_{j,k}$ can be defined. Essentially, a new set of vector functions, $\mathbf{p}_{j,k}(u, v)$, is defined in the following manner

$$\mathbf{p}_{j,k}(u, v) = \sum_{q=0}^{3} \sum_{r=0}^{3} u^a v^r \mathbf{R}_{q,r} \tag{8.24}$$

The parameters u and v vary from 0 to 1 in the directions illustrated in the figure. The similarity between the above equation and equation (8.12) should be noted.

Given the nodal points, $p_{j,k}$, it is necessary first to determine the functions $\mathbf{R}_{q,r}$ and then solve for the tangent vectors, $\mathbf{t}_{j,k}$ and $\mathbf{s}_{j,k}$. To determine expressions for $\mathbf{R}_{q,r}$ we follow the same slope and parameter matching approach that was used for a single space curve. However, in this case all properties must be matched for each row and each column of points in the array. This procedure results in the expressions for $\mathbf{R}_{j,k}$ (i.e., R_{00} to R_{33}) that can be substituted into equation (8.24) to yield an expression for any point on the surface in terms of u, v, $\mathbf{p}_{j,k}$, $\mathbf{t}_{j,k}$ and $\mathbf{s}_{j,k}$. The unknown tangent vectors are determined using matrix methods.

To provide sculptured surface data to NC equipment, interpolation must be possible at any point on the surface and an offset surface must be obtained. The offset surface is a locus of points *parallel* to the original free-form surface. The center of a ball-end cutting tool, tangent to the sculptured surface, would lie on the offset surface.

Interpolation of a point on a free-form surface is accomplished using equation (8.24) and the parameters u, v. Referring to Figure 8.17, u and v are selected so that the appropriate interpolated point is defined. Each component of $\mathbf{p}_{j,k}(u, v)$, that is, x, y, and z can then be calculated. Similarly, the tangent

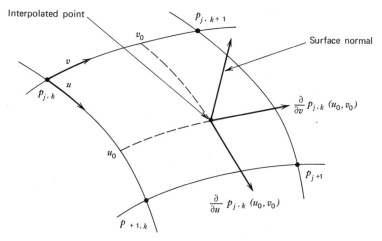

Figure 8.17 Interpolation and normal calculations for 3-D surfaces.

vectors at the interpolated point (u_0, v_0) are

$$\frac{\partial}{\partial u}\, \mathbf{p}_{j,k}(u_0, v_0)$$

and

$$\frac{\partial}{\partial v}\, \mathbf{p}_{j,k}(u_0, v_0)$$

Hence, by computing the cross product of the tangent vectors, a vector normal to the surface is defined. These steps, implemented using computer methods, are used to provide cutter coordinates for NC machines.

8.5.2 The Bi-Cubic Surface Patch

A sculptured surface may also be constructed by building sets of surface *patches* that are expressed in terms of parametric relationships. The analytic model of the surface patch, generally attributed to S. A. Coons (Reference 13), was initially developed for interactive computer-aided design applications. Nevertheless, patch techniques have been adapted to NC applications and are now used almost exclusively in the definition of sculptured surfaces for numerical control.

The advantages of patch-type sculptured surfaces are flexibility and the ease with which surface analysis can be performed in matrix form. The free-form surface is constructed using patches which are described in terms of *patch boundaries* and *blending functions*. The patch boundaries are generally defined by four curves (although three may be used in a degenerate case) which enclose a given surface area. Blending functions are used to maintain slope continuity between adjacent surface patches.

The analysis of a surface patch begins with a surface point vector, $\mathbf{V}(u, w)$, which is defined by a matrix equation of the form*

$$\mathbf{V}(u, w) = [\mathbf{U}][\mathbf{M}][\mathbf{B}][\mathbf{M}]^{\mathrm{T}}[\mathbf{W}]^{\mathrm{T}} \qquad (8.25)$$

The U and W matrices represent the parameters u and w which vary from 0 to 1.

$$[\mathbf{U}] = [u^3 \quad u^2 \quad u \quad 1]$$
$$[\mathbf{W}] = [w^3 \quad w^2 \quad w \quad 1]$$

The cubic nature of each parameter should be noted. **M** is a coefficient matrix

---

* The superscript **T** indicates the transpose of the referenced matrix; i.e., all rows and columns are exchanged.

that is derived by considering a single curve along the patch boundary and is shown to be

$$\mathbf{M} = \begin{bmatrix} 2 & -2 & 1 & 1 \\ -3 & 3 & -2 & -1 \\ 0 & 0 & 1 & 0 \\ 1 & 0 & 0 & 0 \end{bmatrix}$$

Finally, the **B** matrix, or boundary condition matrix, contains all the information concerning patch boundaries. The matrix consists of sixteen terms divided into four submatrices as shown below

$$[\mathbf{B}] = \begin{bmatrix} [\mathbf{P}] & [\mathbf{P}_w'] \\ [\mathbf{P}_u'] & [\mathbf{T}_{uw}] \end{bmatrix}$$

where the **P** matrix contains the four corner coordinates, $\mathbf{P}_w'$ and $\mathbf{P}_u'$ contain the corner slopes with respect to w and u, respectively, and $\mathbf{T}_{uw}$ is the *twist vector matrix* containing corner cross derivatives.

To clarify the above description of equation (8.25), consider the surface patch illustrated in Figure 8.18. Along a given boundary curve the function $V(u, w)$ takes the form of a cubic curve. For example, along curve 1

$$V(u, 0) = Au^3 + Bu^2 + Cu + D \tag{8.26}$$

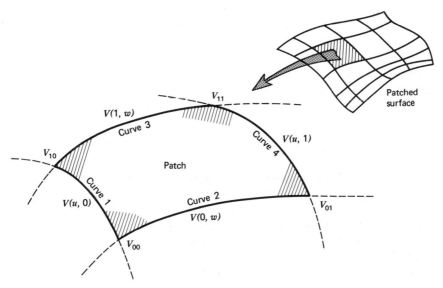

Figure 8.18 Coon's surface patch (Reference 13).

where A, B, C, and D are coefficients. Similar to the Ferguson development, the coefficients of equation (8.26) are obtained by considering slope and position continuity across patch boundaries.

As illustrated in Figure 8.18, we define a shorthand notation for all u and w in the form

$$V_{10} = V(1,0)$$

also the slopes,

$$V_{10u} = \frac{\partial}{\partial u} V(1,0)$$

and twist vectors,

$$V_{10uw} = \frac{\partial^2}{\partial u \, \partial w} V(1,0)$$

In general, equation (8.25) can be rewritten in the form

$$V_{(u,w)} = [u^3 \; u^2 \; u \; 1][M] \begin{bmatrix} V_{00} & V_{01} & V_{00w} & V_{01w} \\ V_{10} & V_{11} & V_{10w} & V_{11w} \\ V_{00u} & V_{01u} & V_{00uw} & V_{01uw} \\ V_{10u} & V_{11u} & V_{10uw} & V_{11uw} \end{bmatrix} [M]^T \begin{bmatrix} w^3 \\ w^2 \\ w \\ 1 \end{bmatrix} \qquad (8.27a)$$

For any patch, the matrix product $[M][B][M]^T$ is a constant matrix which is generally written $[S]$. Hence, from equation (8.25)

$$V(u,w) = [u^3 \; u^2 \; u \; 1][S] \begin{bmatrix} w^3 \\ w^2 \\ w \\ 1 \end{bmatrix} \qquad (8.27b)$$

The elements of S can be obtained from given coordinates and slope continuity considerations. Hence, any point on the patch is defined using equation (8.27b). In actuality, equation (8.27b) becomes three equations—one for each coordinate, x, y, and z.

8.5.3 An Example of Calculation of a Matrix **M**

The boundary curve of a surface patch takes the cubic form expressed in equation (8.26). Using the equation as a starting point, the elements of a matrix M, that relate position and slope to the cubic function to the coefficients A, B, C, and D, can be calculated.

Along a boundary curve in the u direction

$$F(u) = Au^3 + Bu^2 + Cu + D$$

or in matrix notation

$$F(u) = [u^3 \ u^2 \ u \ 1] [A \ B \ C \ D]^T$$

and the slope

$$\frac{\partial}{\partial u} F(u) = F'(u) = [3u^2 \ 2u \ 1 \ 0] [A \ B \ C \ D]^T$$

If we consider the starting and end points of the curve (i.e., $u = 0$ and $u = 1$),

$$F(0) = D$$
$$F(1) = A + B + C + B$$
$$F'(0) = C$$
$$F'(1) = 3A + 2B + C$$

or

$$\begin{bmatrix} F(0) \\ F(1) \\ F'(0) \\ F'(1) \end{bmatrix} = \begin{bmatrix} 0 & 0 & 0 & 1 \\ 1 & 1 & 1 & 1 \\ 0 & 0 & 1 & 0 \\ 3 & 2 & 1 & 0 \end{bmatrix} \begin{bmatrix} A \\ B \\ C \\ D \end{bmatrix} = [\mathbf{m}][\mathbf{c}]$$

If we define $[\mathbf{M}] = [\mathbf{m}]^{-1}$ then

$$[A \quad B \quad C \quad D]^T = [\mathbf{M}][F(0) \quad F(1) \quad F'(0) \quad F'(1)]$$

Using standard matrix procedures

$$[\mathbf{M}] = [\mathbf{m}]^{-1} = \begin{bmatrix} 2 & -2 & 1 & 1 \\ -3 & 3 & -2 & -1 \\ 0 & 0 & 1 & 0 \\ 1 & 0 & 0 & 0 \end{bmatrix}$$

8.5.4 NC Implementation of Sculptured Surfaces

Implementation of sculptured surface techniques within NC part programming languages has been accomplished with varying degrees of success. Surface definition and evaluation are available in modified versions of APT and other special purpose part programming systems in the industry.* Each system maintains its own set of language conventions and descriptive requirements.

The sculptured surface extension to the APT system (Reference 15) provides a representative example. In this system the free-form surface is defined by a

---

* Probably the best known industrially developed systems are Boeing's APTLFT and FMILL, and Douglas Aircraft's DAC system. The major auto makers, Ford and General Motors in the United States and Renault in France, have also developed sophisticated free-form surface techniques.

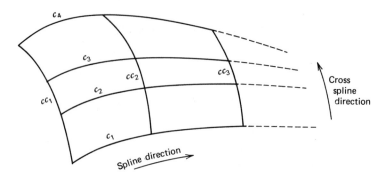

Figure 8.19   Sculptured surfaces in the APT language (Reference 15).

mesh of patches. Initially, the surface is described by a series of *splines* and *cross splines* as illustrated in Figure 8.19. From this original description, the arithmetic element interpolates a patch structure. A language geometry statement takes the form

$$surface = \text{SSURF/GENCUR}, c_1, c_2, c_3, c_4, \text{CRSSPL}, cc_1, cc_2, \ldots$$

where data values are illustrated in the Figure. The three-dimensional spline curves have been previously defined in the APT program.

Other surface definitions are also possible. The sculptured surface may be defined using a *mesh* of predefined points or a combination of subsurfaces.

Recent developments in NC applications of the sculptured surface include: (1) methods that allow the generation of arbitrary surfaces of revolution, (2) the implementation of free-form check surfaces, (3) removal of surface *defects* by data smoothing, and (4) definition of a *synthesized surface* which is a combination of free-form and analytic surfaces. As these processes become formalized, powerful new analysis techniques will be available to the NC part programmer.

References

1. Wylie, C. R., *Advanced Engineering Mathematics*, 4th ed., McGraw-Hill, New York, 1975.
2. Shinners, S. M., *Modern Control System Theory and Application*, Addison-Wesley, Reading, Mass., 1972.
3. Feldman, C. G., "A General Description of the New APT System Arithmetic Element," *Investigations in Computer-Aid Design*, Massachusetts Institute of Technology, Report 8436-IR-1, January 1961.
4. *System/360 APT Numerical Control Processor, Version 4, System Manual*, IBM, manual GY20-0080, February 1969.

5. Hildebrand, F. B., *Introduction to Numerical Analysis*, 2nd ed., McGraw-Hill, New York, 1974.

6. Ferguson, J. C., "Multivariable Curve Interpolation," *Journal of the ACM*, April 1964.

7. Ahuja, D. V., "An Algorithm for Generating Spline-like Curves," *IBM Systems Journal*, vol. 7, nos. 3, 4, 1968.

8. Akima, H., "A New Method of Interpolation and Smooth Curve Fitting Based on Local Procedures," *Journal of the ACM*, October 1970.

9. Fuller, L. E., *Linear Algebra with Applications*, Dickenson Publishing Co., Belmont, Calif., 1966.

10. Ayres, F., *Matrices*, McGraw-Hill (Schaum's Outline Series), New York, 1962.

11. Lane, E. R., *Metric Differential Geometry of Curves and Surfaces*, The University of Chicago Press, 1940.

12. Birkhoff, G., and Garabedian, A., "Smooth Surface Interpolation," *Journal of Mathematics and Physics*, vol. 39, no. 4, December 1960.

13. Coons, S. A., *Surfaces for Computer Aided Design of Space Forms*, Project MAC, Massachusetts Institute of Technology, MAC-TR-41, June 1967.

14. Bezier, P., *Numerical Control—Mathematics and Applications*, Wiley, New York, 1972.

15. Hinds, J. K., "Part Programming Language for the Definition and Use of Synthetic (or Sculptured) Surfaces," (APT SSX3), IIT Research Institute, March 1972.

Additional sources of information

Khol, R., "The Search for the Sculptured Surface," *Machine Design*, vol. 45, March, 22, 1973, pp. 154 ff.

Almond, D. B., "Numerical Control for Machining Complex Surfaces," *IBM Systems Journal*, vol. 11, no. 2, 1972.

Peters, G. J., "Parametric Bi-cubic Surfaces," a paper presented to the Society of Industrial and Applied Mathematics, National Meeting, June 1973.

Olesten, N. O., *Numerical Control*, Wiley-Interscience, New York, 1970.

Problems

For Problems 1–3, use analytic geometry to derive expressions that enable the specification of all cut vector end points required for NC input. In all cases, assume the cutting tool has a radius, r, and that lines are defined using slope-intercept form (i.e., $y = mx + b$).

1. A cutting tool is centered at point (x_0, y_0) and moves along a line, l_1, which is tangent to a circle with radius R. The cut vector is to terminate at the point of tangency between the cutter periphery and the circle.

2. A cutting tool is centered at a point (x_1, y_1) which makes the tool periphery tangent to a circle, c_1, of radius R_1, centered at (X_{c1}, Y_{c1}). Another circle, c_2, of radius R_2, is centered at (X_{c2}, Y_{c2}), where $Y_{c2} = Y_{c1}$ and $|X_{c2} - X_{c1}| < (R_1 + R_2)$. The cutting tool

is to move along the circumference of c_1 until it is tangent to c_2. Develop an expression for this tangency point only.

3. A cutting tool is centered at point (x_0, y_0) and moves along a line, l_1, which intersects an ellipse. The ellipse is centered at the origin and has a major axis $= a$, and a minor axis $= b$. The cut vector is to terminate so that the cutting tool center lies on the ellipse.

4. A parabolic arc, centered at the origin, is defined by the equation $y^2 = 10x$ in the range $0 \le x \le 5$. Calculate a cutter path with line segments no longer than 1 unit which will move a cutting tool of radius $= 1$ tangent to the arc on its convex side.

5. A cutting tool center point moves along a line, l_1, which passes through the points $(10, 8, 6)$ and $(-8, -6, -6)$. Determine the end point of a cut vector which begins at the first point moving along l_1 until the cutter is tangent to a sphere of radius $R = 2$, centered at the origin. The cutter radius $r = 1$.

6. Derive expressions similar to equations (8.6) for secantial and tangential methods of arc approximation. What are the benefits and drawbacks of the three approximation methods?

7. Derive expressions for the tolerance angle, ϕ, when a cutter of radius r travels on the concave side of a circular arc of radius R. Plot ϕ versus tolerance value, t, for $\Delta R \approx R$ (i.e., $r \ll R$); for $\Delta R = 0.5R$. What would happen if $\Delta r = 0$?

8. Develop an iterative procedure that could be used to maintain tolerance on higher order curves (e.g., cubic curves). Outline your approach.

9. Use the pull vector calculation procedure to determine the minimum distance between a parabola $y^2 = 20x$ and a circle $(x + 6)^2 + (y - 3)^2 = 16$. Draw the curves and illustrate your results graphically.

10. Explain why a parametric spline can accommodate a curve with an infinite slope whereas a mathematical spline cannot. Why are cubic curves chosen instead of, say, second degree curves?

11. In your own words, explain boundary condition continuity for a parametric spline. Starting with equations (8.15), show that

$$\mathbf{t}_j + 4\mathbf{t}_{j+1} + \mathbf{t}_{j+2} = 3(\mathbf{p}_{j+1} - \mathbf{p}_j)$$

12. (a) Develop a computer program that will calculate the tangent vectors for a parametric spline. Use the Ferguson approach outlined in the text.
(b) Develop routines for interpolation and normal calculation.
Note: Matrix inversion subroutines are generally available at most computer installations.

13. Based on the general parametric spline configurations, does nominal point (i.e., knot) spacing have an effect on the resulting spline? Explain your answer in terms of tangent vector calculation procedures.

14. Use the Ferguson approach to compute the data necessary to spline fit the following set of points in two dimensions:

x	y
2.0	3.0
10.0	8.0
19.0	8.0

Based on the computed data, draw a line segment approximation of the curve such that each segment is approximately 2 units long.

15. From the curve developed in Problem 14, draw an offset curve at a distance 3 units *above* the original curve.

16. Is every point on an offset spline exactly the same distance from the original? Discuss your answer based on the number of offset points used to define the *parallel* spline and the local curvature of the original spline.

17. Use the Ferguson approach to compute the tangent vectors for the following set of points, defined in three dimensions:

x	y	z
0	0	0
3	6	4
-1	8	7

Sketch the curve in three orthogonal views.

18. An analytic surface is defined by the equation:

$$\frac{x^2}{16} - \frac{y^2}{8} + \frac{z^2}{4} = 1$$

Sketch the surface, and calculate the tangent and normal vectors at the point $(4, 2, 1.414)$. The surface is bounded by the planes $y = \pm 3$.

19. An ellipsoid is defined by the equation:

$$\frac{x^2}{a^2} + \frac{y^2}{b^2} + \frac{z^2}{c^2} = 1$$

Find general expressions for the tangent vector and normal vector for any point x, y, z.

20. State the continuity conditions at the point, $\mathbf{p}_{j,k}$, that must be used to solve for $\mathbf{R}_{q,r}$ in question (8.24). Use Figure 8.16 as a guide. What are the conditions at the boundary of the surface?

21. In order to machine a sculptured surface, the cutting tool must move across the surface from patch to patch. That is, a curvilinear progression of points must be interpolated. Assuming that the cutting tool begins at an arbitrary boundary and moves across the surface in an arbitrary direction, outline a method which would (1) select which patch to consider as motion commences and continues and (2) calculate the curve coordinates from which surface normals and the resulting cutter center points could be calculated.

Chapter Nine
Process Optimization

Optimization processes have been developed to improve the operational characteristics of NC machine tool systems. Two distinct methods of optimization are *adaptive control* and *machinability data prediction*. Although both techniques have been developed for metal cutting operations, adaptive control finds application in other technological fields.

The adaptive control (abbreviated AC) system is an evolutionary outgrowth of numerical control. AC optimizes an NC process by sensing and logically evaluating variables that are not controlled by position and velocity feedback loops. Essentially, an adaptive control system monitors *process variables*, such as cutting forces, tool temperatures, or motor torque, and alters the NC commands so that optimal metal removal or safety conditions are maintained.

Machinability data is essential for the successful application of NC to machining.

9.1 Adaptive Control

A typical NC configuration (Figure 9.1a) monitors position and velocity output of the servo system, using feedback data to compensate for errors between command and response. NC feedback is directly related to command data. That is, both position and velocity information are explicitly specified on NC input media. Once cutter path and feedrate have been input, the NC system attempts to satisfy programmed commands without regard to the actual process being performed. Adaptive control introduces an additional feedback loop to the NC system.

The AC feedback loop (Figure 9.1b) provides sensory information on other *process variables* such as: workpiece–tool air gaps, material property variations, wear, cutting depth variations, or tool deflection. This information is determined by techniques such as monitoring forces on the cutting tool, motor torque variations, or tool–workpiece temperatures. The data is processed by an

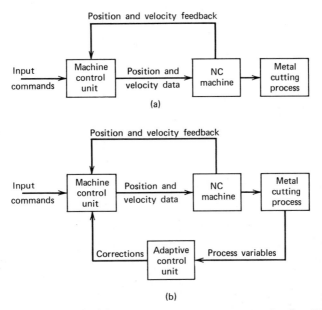

Figure 9.1 Schematic diagrams for conventional and adaptive NC systems.

adaptive controller that converts the process information into feedback data to be incorporated into the MCU output.

9.1.1 Optimizing Adaptive Control

An optimizing adaptive control system manipulates input variables in a manner that optimizes a set of measurable process variables. Figure 9.2 illustrates such a system. A process controller transmits a number of controllable input variables, X_i, that provide command data for the process. Other inputs, in the form of system disturbances, D_i, can have a significant effect on system output, but cannot be directly controlled. Measurable output variables, P_j, also called process variables, are monitored and passed to a *performance computer*, the first element in an AC system.

As indicated in Figure 9.2, the input to the performance computer consists of X_i, P_j, and a set of desired performance criteria, C_k. Using predefined functional relationships of the form

$$M = g(X_i, P_j) \tag{9.1}$$

the performance computer uses optimization logic to compute the best

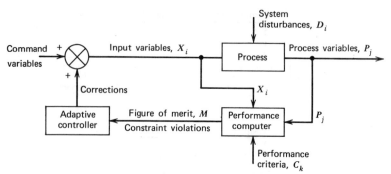

Figure 9.2 Optimizing adaptive control.

combination of input variables, X_i, based on C_k and instantaneous values of P_j. The value M in equation (9.1) is termed the *figure-of-merit*. _merit function ?_

Output from the performance computer is passed to the adaptive control unit, and this generates corrections to X_i based on M. However, the mechanics of every process require that X_i remain within reasonable bounds. For this reason constraints are placed on the corrected input variables so that the physical limits of the process are not violated. When the performance computer dictates a correction that places X_i outside established bounds, a *constraint violation* occurs and the input variable correction is modified.

9.1.2 AC-NC Systems

Optimizing adaptive control systems can be applied to any process with discrete, measurable input and output. Conventional NC equipment cannot easily adapt to changes in its processing environment. That is, the NC machine is *programmed* to execute a specific set of commands, and the MCU cannot adjust the part program commands *on the fly.*

Although the MCU maintains control over position and velocity commands, an adaptive control unit senses changes in the processing environment and adjusts machine tool functions in a manner that optimizes overall system operation. By coupling precise programmed control with instantaneous adaptive capabilities it is possible to achieve machine tool and workpiece protection, simplified part programming, improved dimensional control, and overall process optimization. In addition, the system is less dependent upon the machine operator.

To illustrate how these benefits are realized, consider a simple adaptive NC system that is capable of sensing forces on the cutting tool. A system schematic is illustrated in Figure 9.3. The ability to monitor cutting forces on the tool

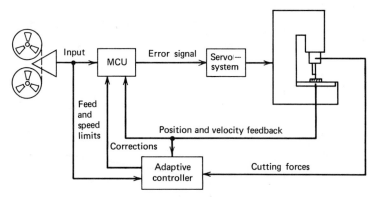

Figure 9.3 A simplified adaptive NC system.

provides implicit information on tool deflection and evaluative information regarding tool wear, efficacy of the programmed feedrate, and variation in material properties of the workpiece.

Both the machine tool and the workpiece are protected by adjusting feedrate and spindle speed so that cutting forces do not exceed prescribed values. In conventional NC systems, there is always the danger that feedrates or speeds will be programmed at values that lead to tool breakage or excessive tool wear. AC automatically adjusts feedrate and/or speed when it senses tool forces outside of prescribed bounds.

In optimizing AC systems, programmed feeds and speeds are eliminated. The part programmer need only specify appropriate bounds for feedrate and speed, and the adaptive controller adjusts the actual values based on process feedback information. Hence, NC part programming is simplified and performance is optimized.

Although servo system accuracy is extremely important relative to dimensional accuracy, tool wear and workpiece-tool deflection also cause errors that are not compensated by the NC system. Dimensional error associated with tool deflection is illustrated in Figure 9.4. High cutting forces cause tool deflection so that the programmed tool location, which assumes no deflection, differs from the actual tool location. A dimensional error results* regardless of the servo system accuracy.

By sensing tool cutting force, F, the AC system maintains feedrates that keep deflection within acceptable levels. This technique also reduces vibration and improves workpiece surface finish.

---

* It should be noted that tool deflections are generally of the order of 0.02 mm or less. However, high accuracy operations require that deflection be minimized.

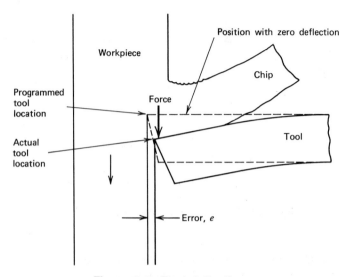

Figure 9.4   Tool deflection.

Conventional NC operations rely on the part programmer and the machine operator to insure that safe conditions are maintained. To protect an expensive NC machine, the operator often overrides the feedrate and speed commands using experience judgements. Although such intervention is sometimes justified in emergency situations, a gross reduction of production rate over the whole process time span results in inefficient operation. By monitoring tool force on a continuous basis, production rate is reduced only when monitored force exceeds safe levels.

Adaptive control serves to improve the machining process by maximizing the volume of metal removed. Tests by the Bendix Corporation (Reference 1) have shown that an adaptive NC system can reduce production times by as much as 30 percent when compared to conventional NC. Time, however, is not the only parameter that can be optimized. By redefining optimization logic, other parameters such as surface finish or tool life can be improved.

9.2 Elements of an Adaptive NC System

The metal cutting process is affected by many interrelated variables, and in addition to cutting forces and motion feedback, adaptive NC systems monitor spindle horsepower, cutter torque, tool–workpiece temperature, and tool vibration. These systems require a sophisticated array of transducers and

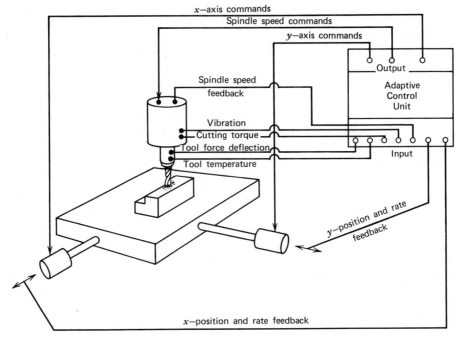

x—axis commands
Spindle speed commands
y—axis commands
Output
Adaptive
Control
Unit
Spindle speed
feedback
Vibration
Cutting torque
Tool force deflection
Input
Tool temperature
y—position and rate feedback
x—position and rate feedback

Figure 9.5 Adaptive NC system configuration.

sensing devices so that process data can be generated and passed to the adaptive controller.

For purposes of clarity, the adaptive controller and the MCU are illustrated as separate elements in the NC-AC control loop. Although each element performs different functions and logical operations, they are housed within a single cabinet and are collectively termed the *adaptive control unit* (ACU). Figure 9.5 illustrates ACU inputs and outputs of both motion and process data.

Cutter force, tool deflection, and torque are sensed using strain gages mounted near the cutting tool or on the spindle, whereas motor horsepower input is determined by measuring motor current flow. Using the thermocouple principle, the voltage produced at the tool and workpiece junction is a measure of the tool-chip interface temperature. Tool vibration is determined by mounting an accelerometer on the spindle housing.

9.2.1 Constraint Correction Mode

An adaptive control system continuously attempts to increase performance in a selected parameter until a *constraint violation* occurs. Hence, AC has two modes of operation, *optimization* and *constraint correction*, that are invoked as a function of the current state of the process variables.

Constraints define the permissible range of operation for the ACU. The performance computer contains constraint violation logic that can detect when either motion or process variables exceed predefined bounds. For example, the AC system in Figure 9.5 might have the following constraints: a maximum/-minimum for spindle speed, and maximums for feedrate, tool force, temperature, vibration, and torque loading. Should any of these bounds be exceeded, a constraint violation signal would be issued to the ACU logic subsystem illustrated in Figure 9.6.

Depending on the process variable(s) that caused the constraint violation, the logic subsystem increments or decrements the value of the ACU outputs in a manner that returns the process variable(s) to prescribed bounds. For example, if maximum tool force has been exceeded, feedrate and/or spindle speed might be reduced. Because it is possible that correction of one constraint violation might cause a different variable to exceed predefined bounds, the ACU subsystem logic is programmed so that the AC system does not oscillate between constraint violations.

The constraint violation detection system is also used to shut down opera-

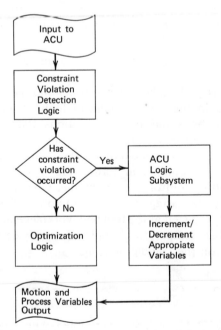

Figure 9.6 ACU logic and constraint violation handling.

tions when dangerous conditions are detected. For example, machining operations are stopped immediately when sudden temperature increase indicates a
breakdown of the tool cutting edge. Another ACU logical function is tool life
monitoring. The expected tool life (in minutes) is input to the ACU through the
NC communication media. The ACU keeps track of the remaining time for
each tool and produces a warning signal when the end of tool life expectancy
has been reached.

9.2.2 Optimization Mode

In the optimization mode the ACU maximizes the overall efficiency of the
production process or selected process parameters. The figure-of-merit, M,
sometimes called the utility function or *merit function*, is a numerical measure
of efficiency. The figure-of-merit is a number whose magnitude indicates the
merit or desirability of a given combination of variables (Reference 2). To
optimize the machining process, one or more figures-of-merit, representing
various interrelationships between process and motion variables, are maxim-
ized without causing a constraint violation.

The optimization process is an n-dimensional maximization problem, where
n is the number of independent variables that comprise the merit function. To
illustrate the maximization procedure, consider a two-dimensional (i.e., $n = 2$)
definition for M, where the independent variables are feedrate, f, and spindle
speed, s. Therefore

$$M = g(f, s) \qquad (9.2)$$

where g represents a function relating f and s to the figure-of-merit, M.

Conceptually, it is possible to visualize the merit function as a three-
dimensional response surface, consisting of a two-dimensional parameter plane
and a third axis along which the figure-of-merit is measured. This surface takes
the form of a contour map, illustrated in Figure 9.7 (Reference 3).

To maximize M, a *hill-climbing* optimization technique can be used. The
response surface is ascended in the direction of steepest slope. The local
gradient at the starting point is determined, and the operating point (i.e., the
process variable combination) is altered so that a new operating point is
established. At each new operating point, the local gradient is determined and
the next operating point is calculated one step closer to the maximum. This
procedure is continued until the peak of the hill, when the slope is zero, is
reached.

A wide variety of optimizing techniques are applied to merit function
response surfaces. Two important procedures applicable to expression (9.2) are
the *steepest ascent procedure* (Reference 3) and the *sequential simplex search*
technique (Reference 4).

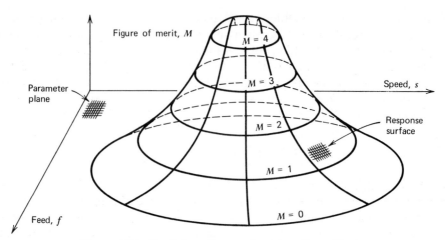

Figure 9.7   Merit function surface for $M = g(f, s)$.

9.2.3 The Steepest Ascent Procedure

For the merit function described in expression (9.2), the steepest ascent
procedure begins at a point that represents the current set of operating
conditions. The feedrate is increased by a small increment, Δf, in the positive
direction, and the merit function change, ΔM_1, is recorded. The speed is then
changed by a small increment in the positive direction, and the corresponding
merit function change, ΔM_2, is recorded. The values of ΔM_1 and ΔM_2
approximate the partial derivative of M with respect to f or s, respectively.
Therefore, when the next move changes f by $k \Delta M_1$ and s by $k \Delta M_2$, movement
along the gradient has been approximated. The value k is a scale factor that
defines the step size for movement *up* the hill. The procedure, illustrated in
Figure 9.8a, is repeated for each operating point. The steepest ascent procedure
can also be applied when M is defined by more than two independent process
variables.

9.2.4 The Sequential Simplex Search

The sequential simplex search technique uses a geometric figure, called a
simplex to aid hill climbing. For a merit function defined by two process
parameters, the simplex is an equilateral triangle. The hill climbing method is
illustrated in Figure 9.8b. Beginning at point p_1, two additional merit function
values are calculated choosing f and s so that p_1, p_2, and p_3 form an equilateral
triangle, the simplex. The basic simplex technique rejects the *least optimum*

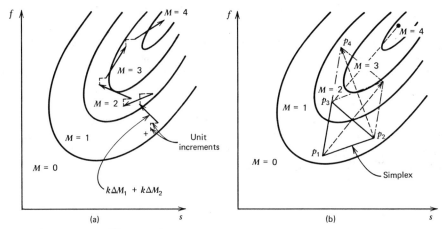

Figure 9.8 Optimization techniques for *hill-climbing*. (a) Steepest ascent procedure. (b) Sequential simplex search.

point and moves away from it in a direction that creates a new simplex, comprised of points p_2, p_3, and p_4. By continually constructing new simplexes, the path eventually leads to an optimum merit function value. Using this technique, more than two process parameters can be evaluated. For example, three parameters can be analyzed by defining a tetrahedron simplex.

The simplex search stalls when the merit function response surface changes contour rapidly; that is, when two equally nonoptimum values cause the search to oscillate. Therefore, two additional rules are applied:

1. A return to the point that has just been left is invalid, and movement goes to the next highest rejected value.
2. The search is terminated after a specified number of iterations. This reduces oscillation about the maximum.

9.3 Sensing Methods for Adaptive Control

An adaptive NC system requires current and accurate process data. Unlike position and velocity transducers, AC instrumentation measures process variables indirectly. Temperature is measured as a voltage difference; forces are measured in terms of displacement; and horsepower is related to motor current. The prime requisites for AC instrumentation are *sensitivity* and *stability*. The sensing device should be capable of detecting process parameter changes that affect machining conditions. It must provide a stable output over a range of specified values.

9.3.1 AC Instrumentation

Although many variables affect changes in the metal cutting process, cutting force is the easiest to measure, and for this reason nearly all* adaptive control systems monitor forces. The most common device used for the measurement of cutting force is the *strain gage*. Different types of strain gages are illustrated in Figure 9.9.

The strain gage is comprised of a small flat electrical conductor bonded to the surface of the tool or support structure. As the structure is deflected due to cutting forces, the strain gage is either stretched or compressed. The electrical resistance, R_g, of the strain gage is written:

$$R_g = \frac{\rho L}{A}$$

where L and A are the conductor length and cross-sectional area, and ρ is the *resistivity* of the conducting material. It can be shown (Reference 5) that a change in length (due to deformation) will change the resistance of the gage. A typical strain gage configuration is illustrated in Figure 9.10.

The strain gages are bonded to the tool holding structure so that both horizontal and vertical forces cause corresponding tool strains which can be measured. The gage resistances form part of a bridge circuit that allows ΔR_g to be measured in terms of current flow for an applied voltage. When calibrated, a relationship between strain, ϵ, and current, I, or voltage for the entire range of operating values is developed. Furthermore, known structural dimensions, stiffness, and gage linearity enable ϵ to be directly converted to deflection or force.†

* In some AC systems only tool life is considered. For such configurations, tool temperature may be the primary consideration.
† Strain gage techniques are presented in engineering measurement texts (e.g., References 5–7).

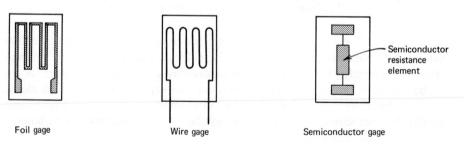

Foil gage   Wire gage   Semiconductor gage

Figure 9.9 Typical strain gage configurations.

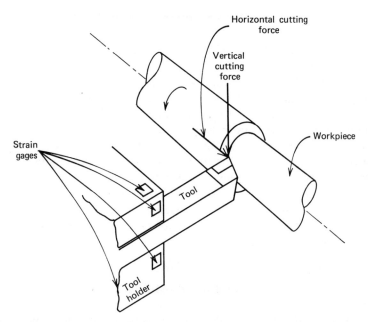

Figure 9.10   Strain gage placement for force measurement.

To illustrate the use of strain gages, consider a force sensing arrangement that monitors cutting force using a work holding base (Reference 1). The base is secured to the machine slide by four bolts instrumented with strain gages to measure the axial force on each bolt. The algebraic sum of the current flow provides a resultant signal that is proportional to the tangential tool force. In the system described in Reference 1, a cutting force of 5000 pounds provides an output voltage of 10 volts with linearity and repeatability of 0.1 percent full scale. This is transmitted to the ACU in digital form and serves as a process variable in the adaptive control strategy.

Both *current* and *temperature sensing devices* may be attached to the spindle drive motor to monitor motor work. These sensors guard against motor overload by monitoring motor temperature and input current. When the temperature sensing devices indicate that the temperature of the motor is excessive, feedrate is reduced until the motor cools to a safe level.

The temperature at the tool-workpiece interface is an important indication of tool wear and cutting efficiency. The most effective means of sensing machining temperature is the use of the cutting tool and the workpiece metal as two elements of a *thermocouple*. The *emf* produced between these two dissimilar metals provides data for the overall ACU evaluation of cutting conditions.

9.3.2 Air Gap and Interrupted Machining

Instrumentation enables the ACU to determine machine tool cutting mode on a continuous basis. Three possible cutting modes are encountered during machining. The first, called *air gap*, occurs when the tool is not in contact with the workpiece for more than one revolution of the cutting tool. *Interrupted cutting*, such as that produced by a milling cutter, is the second mode; and *continuous cutting*, as occurs in lathe turning, is the third.

When the ACU recognizes an air gap condition—that is, tool force sensors indicate zero force readings—the maximum feedrate per spindle revolution is requested. This maximum requested feedrate is determined by the safe impact speed when tool and work meet and machining commences. Recognition of air gap enables production rate to be increased with no loss of machining quality. Feedrate data for interrupted and continuous cutting conditions is based upon machinability considerations.

9.4 Machinability*

The specification of suitable feeds and speeds is essential in conventional and NC cutting operations. *Machinability data* is used to aid in the selection of metal cutting parameters based on the machining operation, the tool and workpiece material, and one or more production criteria.

The selection of feedrate and speed is the responsibility of the manufacturing engineer who must communicate his requirements to the NC part programmer. Machinability data is chosen so that one or more of the following production criteria is satisfied.

1. *Tool life.* The cutting tool lasts for a specified period of time.
2. *Surface finish.* A specified surface smoothness is achieved and maintained.
3. *Accuracy.* Tool deflection and vibration are below a specified maximum.
4. *Power consumption.* Power consumption is maintained below a given level.
5. *Economic criteria.* A maximum production rate or minimum cost per piece is achieved.

9.4.1 Machinability Data and Numerical Control

The physical nature of the metal removal process is the same for both conventional and NC machine tools. Hence, there is little difference in the

---

* Parts of this section are based on a paper (Reference 9) presented by the authors at the Third International Conference on Production Research, Amherst, Massachusetts. Reprinted with permission of Taylor & Francis, Ltd.

content of machinability data used for these modes of production. Techniques used to select machinability data for conventional machines have two important drawbacks in relation to NC applications: data is generally presented in a tabular form that requires manual interpolation, checkout, and subsequent revision, and tests on the machine tool are required to find optimum conditions.

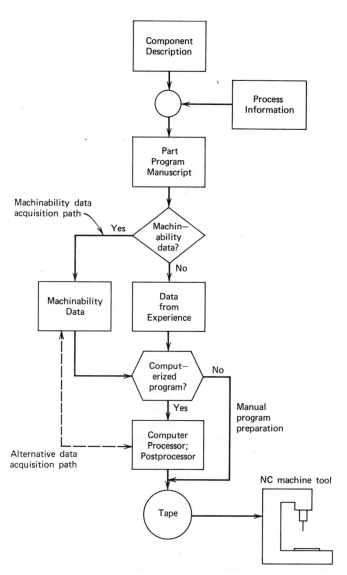

Figure 9.11 Acquisition of machinability data in the NC process flow.

Specialized machinability data systems (Reference 10) have been developed for NC applications to reduce the need for machinability data testing and to decrease expensive NC machining time. Part programming time is also reduced when machinability information is readily available.

A typical process schematic showing the relationship between machinability data and NC process flow is illustrated in Figure 9.11. Both geometric and manufacturing information are used by the part programmer in the preparation of the NC program manuscript. If machinability data is required, a side path (the machinability data acquisition path) into the "machinability" block is taken.

The machinability data obtained from the acquisition path represents varying levels of sophistication. General data may be acquired by using tables found in the appropriate handbook (Reference 11); data may be generated by a single computer program that uses mathematical modeling techniques; or data may be derived from a computerized data base with associated modeling routines. Situations also exist for which the path need not be taken. For example, simple NC operations where improvements in machining performance do not justify the cost of machinability data acquisition. In such cases, past experience is used.

Once machinability data has been acquired, the NC process follows well established paths. If manual programming methods are used, the computer is circumvented and an NC instruction tape is produced directly. For complex workpiece geometries, computer assistance (using a part programming language) may be invoked. The dashed path leading directly from the computer processor to the machinability block represents an important alternative to the original data acquisition path. This path shall be discussed in a later section.

9.4.2 Machinability Data and Adaptive Control

The use of adaptive control significantly reduces the requirement for precise cutting data during the development of an NC part program. AC does not, however, eliminate the need for information that specifies machining constraints, proper cutting tool design, and other factors not considered by the ACU.

An AC system attempts to optimize the machining process with respect to production rate or some other criteria. The merit function description is based on a mathematical model of the cutting process derived from machinability information. Because the merit function optimizes the process relative to one criteria, machinability data can be used if a given component requires some other form of optimization. As an example, consider a situation in which a component must be machined to a precise surface finish, yet the ACU optimizes for maximum production rate. Rather than reprogramming the merit function, machinability data can be used to redefine machining constraints so that both criteria are satisfied.

Hence, machinability data may be used in conjunction with adaptive control to provide additional optimization capabilities. The machinability data acquisition path still remains an important part of the overall AC process.

9.5 Computerized Machinability Systems

Computerized machinability systems make use of empirical expressions and experimentally derived data rather than basic analytical relationships. Because a comprehensive metal cutting theory (i.e., one that enables accurate prediction of process variables over a broad range of operating conditions) has yet to be derived, computerized machinability systems rely on two important techniques:

1. *Mathematical modeling* based on empirically derived equations that best fit a range of experimental data.
2. *Data base systems* that enable machinability data acquisition from a large bank of information.

Both techniques may be used within the same computerized machinability system.

9.5.1 Mathematical Modeling Programs

A mathematical modeling program represents machinability variables in terms of metal cutting parameter groups. In general, a machinability variable, X (e.g., tool life, surface finish), is expressed as

$$X = g(p_1, p_2, p_3, \ldots, p_n)$$

where $p_1, p_2, \ldots, p_n$ are process parameters related to the metal cutting operation. These parameters may be expressed as nondimensional quantities.

One of the earliest mathematical models for the metal cutting process is Taylor's tool life formula (Reference 12) developed in 1907:

$$vT^n = C$$

where v is the cutting speed, T is the length of tool life in minutes, n is an empirically derived *tool life exponent*, and C is a constant of the material. For known values of n and C, cutting speed can be determined based on a specified tool life,

$$\log v = \log C - n \log T \qquad (9.3)$$

Equation (9.3) is a simple model that produces fairly good predictive data over a limited range of conditions. In practice a computerized machinability system often employs complex models that are coupled with sophisticated optimization procedures so that feedrate and speed may be selected for the most economic machining conditions.

9.5.2 Data Base Systems

A computerized machinability data base system makes use of a large *file structure* to aid in the computation of machinability data. A typical system is structured as illustrated in Figure 9.12. Input to the system, which may be processed via batch input (e.g., cards), interactive terminal, or timesharing, is evaluated by a processing program that determines what data base information is required. A typical machinability data base consists of three large information files.

The *material file* contains information concerning the machinability characteristics of the workpiece to be used. Data includes experimentally derived feed and speed information corresponding to various cutting tools and opera-

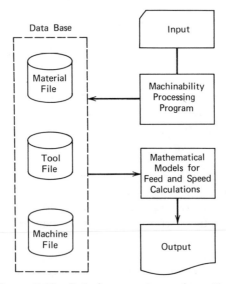

Figure 9.12 Data base system schematic.

tions for a particular workpiece material. The *tool file* describes cutting tools in terms of geometry, that is, diameter, flute length, number of teeth, surface angles; tool holder-adapter configuration; and tool material. The *machine file* contains data defining the operating limitations for a range of machine tools. Maximum and minimum spindle speeds, feedrates, and horsepower are incorporated into the data base. Information regarding machine tool rigidity and accuracy may also be stored.

The machinability data base is generally maintained on high speed storage devices directly accessible to the processing program. Information obtained from the data base is used in analysis modules to form the input for mathematical modeling and interpolation routines. The machinability data base is updated on a regular basis and designed for easy access and maintenance.

9.5.3 Integrated Machinability Systems—EXAPT

Both mathematical modeling programs and data base systems generally operate in a *stand alone* mode. That is, the machinability program is an entity separate from NC language processors and other portions of a computer-aided manufacturing system. An *integrated* machinability system, exemplified by the EXAPT* system (Reference 10), operates as an element of a more complex configuration.

EXAPT is designed to provide a computerized NC part programming system, compatible with the APT language, and a direct interface with predictive machinability analysis procedures. Hence, EXAPT performs two distinct functions. It solves cutter motion geometry so that NC cutter location data may be generated and invokes predictive machinability procedures to determine suitable machining conditions. Both functions are integrated into a single computer system.

A schematic representation of the EXAPT system is shown in Figure 9.13. The part program manuscript for EXAPT contains APT-like statements for geometrical and *technological* features. The geometric processor calculates initial cutter location data based on the workpiece description. The technological processor performs all machinability calculations based on data provided from the EXAPT data base. It also performs cut vector and collision checking calculations to determine final tool path. The output of this element is a standard CL data file that is input to a specific machine tool postprocessor.

The EXAPT language contains additional statements that provide the basic data for machinability calculations. These statements specify operation, materials, and machine tool to be used.

---

*Developed by EXAPT-Verein, Aachen, Germany.

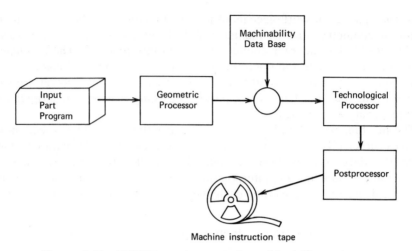

Machine instruction tape

Figure 9.13   EXAPT—an integrated machinability system.

Enhanced versions of APT, such as APT-AC (Reference 13) developed by the IBM Corporation, include a machinability feature. The trend toward integrating machinability analysis and NC part programming will continue as more sophisticated versions of existing NC languages are developed.

References

1. "Adaptive Control," *N/C Handbook*, L. J. Thomas, Ed., 3rd ed., Bendix Corporation, Industrial Controls Division, 1971.
2. Mischke, C. R., "Figures of Merit," *An Introduction to Computer-Aided Design*, Prentice-Hall, Englewood Cliffs, N.J., 1968, Chapter 2.
3. Centner, R. M., "Adaptive Control for Machine Tools," *Numerical Control Today*, Institute of Science and Technology, The University of Michigan, 1967.
4. Leaman, J. M., "A Simple Way to Optimize," *Machine Design*, vol. 45, March 21, 1974, pp. 204–8.
5. Duebelin, E. O., *Measurement Systems; Applications and Design*, McGraw-Hill, New York, 1966, pp. 224–33.
6. Holman, J. P., *Experimental Methods for Engineers*, McGraw-Hill, New York, 1966, pp. 311–38.
7. Perry, C. C., and Lissner, J. R., *The Strain Gage Primer*, 2nd ed., McGraw-Hill, New York, 1962.
8. "Total Automation Approached with Adaptive Control," *Tooling and Production*, vol. 34, no. 6, September 1968, pp. 57–73.
9. Pressman, R. S., and Williams, J. E., "An Investigation of Predictive Machinability

Systems for Numerical Control," *Proceedings of the Third International Confer-ence on Production Research*, Taylor & Francis, Ltd., London, 1975.
10. Hoffman, G. J., "Machinability Data for Optimized NC Production," *N/C Machinability Data Systems*, SME, 1971.
11. *Machining Data Handbook*, 2nd ed., Metcut Research Associates, Inc., 1971.
12. Taylor, F. W., "On the Art of Cutting Metals," *Trans. ASME*, vol. 28, no. 1119, 1907.
13. *System/370 APT-IC and APT-AC Numerical Control Processors, General Infor-mation Manual*, IBM Corporation, publication GH20-1234, 1972.

Problems

1. How will an AC system affect the response time of an NC machine tool? Does the performance computer have a transfer function? Refer to Chapters Two and Three for a discussion of NC system response.

2. Outline the analysis that a designer of the performance computer uses to express deflection in terms of cutting force for a standard lathe tool. State your assump-tions and provide the working equations. What other machining parameters might affect deflection?

3. In this chapter the advantages of adaptive control were discussed. Can you think of a situation in which AC would not be desirable? Would conventional NC be used in such metal cutting operations?

4. If a merit function, M, is defined in terms of three process variables, can the figure-of-merit be expressed graphically? If it can, draw the graphical representation for

$$M = g(x_1, x_2, x_3)$$

   Is it possible to represent M when four process variables are considered? Does such a representation have graphical significance?

5. A response surface for a merit function $M = g(x_1, x_2)$ takes the form of a hemisphere with center at $x_1 = 3$, $x_2 = 3$, and a radius of 4. If contours are chosen at $M = 0$, 0.5, 1, 2, 3, 3.5, draw the contour map indicating minimum and optimum conditions.

6. Using the response surface defined in Problem 5, apply the steepest ascent procedure to trace a path to the optimum. Use Δx_1 and Δx_2 equal to 0.5, and $k = 1$. Show the resulting path, on a contour map, of the response surface. Start the search at $(5, 5)$.

7. Using the response surface defined in Problem 5, use the sequential simplex search procedure to trace a path to the optimum. Let the simplex be an equilateral triangle, 1 unit on each side. Start the search at $(5, 5)$.

8. Using an available measurement text as a guide, discuss the following topics related to strain gage usage:
   (a) Show that a change in length, Δl, changes resistance, ΔRg.
   (b) What is the *gage factor*?

(c) Discuss the bridge circuit used to determine strain.

(d) How is a strain gage *temperature compensated*?

9. Assuming that strain gages are not available, devise another method to determine forces on the cutting tool illustrated in Figure 9.10. Estimate the accuracy of your system. Remember that the ultimate goal is input to the ACU.

10. Sketch a cutting operation in which air gap occurs. If the continuous cut feedrate is f_c, and the feedrate for the air gap is f_a, plot feedrate versus time for continuous cutting, interrupted by an air gap beginning at t_1 and ending at t_2.

11. If a particular component has an optimum continuous cutting feedrate of 0.8 mpm and an air gap feedrate of 2.4 mpm what is the percent increase in production rate between a conventional NC machine and an adaptive NC system for the following conditions: (1) a path 10 m long is to be cut, and (2) for 65% of the path, the cutter moves through air, that is, no metal contact occurs. Assuming machining time cost is \$40/hr, what is the saving for 100 components?

12. Using machinability data available in the literature, plot the general effect of feedrate, depth of cut, rake angle, and workpiece properties on tool life. Use cutting speed as the ordinate of the plot and tool life as the abscissa. A single representative curve should be drawn for each variable.

13. Discuss situations in which machinability data would not be required for an NC metal cutting operation. How is the data supplied in such situations?

14. How do mathematical modeling, data base, and integrated machinability systems differ? How are they the same?

15. Survey recent manufacturing literature for trends in NC optimization processes. Present your results in a form specified by your instructor.

Chapter Ten
Computer-Aided Manufacturing

A computer-aided manufacturing (CAM) system oversees many aspects of manufacture by introducing a hierarchical computer structure to monitor and control various phases of the manufacturing process. Conventional and adaptive NC systems are the predecessors to larger CAM systems. Where NC considers an information feedback loop concerned with a discrete process, CAM develops an integrated information network that monitors a broad spectrum of interrelated tasks and controls each based on an overall management strategy.

Ideally, a CAM system should have three attributes applicable to each phase of the manufacturing process: although supervised, a minimum amount of human intervention should be required for individual process tasks; the system should be flexible and allow processes to be individually programmed; and the CAM system should be integrated, with both engineering design and analysis, using a computer-aided design (CAD) system. Various aspects of this ideal system have been successfully implemented, but much development work continues to be carried out (Reference 1).

10.1 The CAM Hierarchy

A large scale CAM system contains a hierarchical structure of two or three levels of computers that are used to control and monitor individual process tasks. A small (*mini*-) computer is responsible for the management of a single process, with a larger computer monitoring and issuing instructions to a group of small computers. A centralized computer then feeds the system with processed information. This general configuration of a CAM hierarchy is illustrated in Figure 10.1.

A large scale CAM system encompasses three major areas related to the

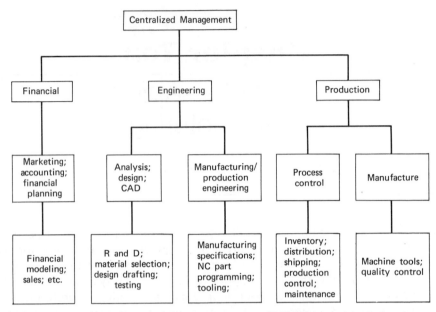

Figure 10.1 Computer-aided manufacturing (CAM) hierarchical structure.

manufacturing process: production management and control; engineering analysis and design; and finance and marketing. Each is comprised of subtasks that are controlled either directly from a large computer (e.g., inventory control), or by a small computer, as in the case of inspection/quality control. Regardless of the control method, the important strength of CAM is that a two-way flow of information occurs.

Because the CAM system oversees many aspects of the manufacturing process, changes dictated by information monitored from one subtask can be translated into control data for some other subtask. For example, in the manufacturing task, machining, inspection, and assembly are all under computer control. When the computer recognizes that a component is continually out of tolerance (based on information feedback from automated testing equipment*), it can be programmed to effect a change in the actual machining process to compensate for the error. Since both subtasks become part of the same information loop, a feedback control system including machining and inspection is established.

* See Chapter 1, Section 1.4.

10.2 NC in Computer-Aided Manufacturing

Although numerically controlled machine tools are essential for the development of operational CAM systems, those described in this text cannot be used in a computer based manufacturing system. Conventional NC and AC machines must be modified so that information may be passed between the MCU and a computer based control system. This modification has resulted in three major developments derived from the NC concept: computer managed numerical control, the *cluster* concept, and new forms of adaptive control.

Computer managed numerical control is a generic term that encompasses DNC (Direct Numerical Control) and CNC (Computerized Numerical Control). DNC and CNC are methods that distribute programmable computing responsibility between a control computer and NC machine tool. Neither system changes the functional characteristics of the NC machine; instead, each provides a means for communication of process data and commands outside the NC machine control loop.

The *cluster concept* is essentially an extension of computer management to more than one kind of machine. A series of machine tools (e.g., those used for milling, boring, grinding) are interconnected by a conveying system that automatically supplies individual machines with components at the required time. Two levels of control and monitoring become necessary. The individual

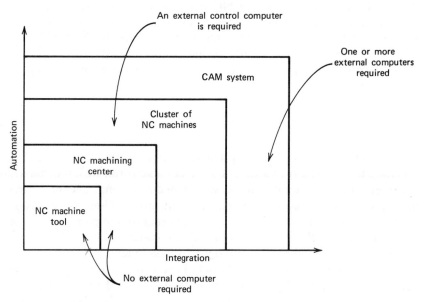

Figure 10.2   Integration and automation for numerical control in CAM.

machines are controlled with computer managed NC, and the cluster itself is managed by a centralized computer coordinating the production output of many clusters. Again, a hierarchical arrangement becomes evident.

Adaptive NC systems are part of the CAM environment. Process information is made available to the centralized computer so that exceptional conditions (e.g., tool breakage) may be detected and corrected. In addition, adaptive feedback may also be recorded and analyzed so that the production efficiency of a given operation may be established.

The use of NC machines in a computer based manufacturing system can be viewed in terms of automation and integration. Referring to Figure 10.2, four levels of manufacturing automation can be defined. The *stand alone* NC machine tool represents an automated operating cycle; an NC machining center automates the entire machining process; a cluster or group of externally controlled NC machines represents a fully automated manufacturing task, and finally, the CAM system itself integrates all lower level methods in an automated manufacturing process.

10.3 Elements of the CAM System

The success of computer-aided manufacturing systems depends upon integration of hardware and software functioning in the overall information flow. CAM hardware elements include NC machine tools, inspection equipment, digital computers, and related devices. CAM software is an interrelated mesh of computer programming systems that serve to monitor, process, and ultimately control the flow of manufacturing data and CAM hardware. In this section the software elements of the CAM system are examined in relation to the flow of information.

10.3.1 The CAM Data Base

A computer based manufacturing system relies upon both *real time* * and stored information that is readily available in an accessible form. Real time data, such as AC feedback, is generally processed immediately and, except for record keeping purposes, can be considered transient information. Stored data provides the CAM system with all necessary input required to perform control and analysis functions. All forms of stored data are maintained in a *data base* that can be accessed at extremely high speed by computer.

The level of complexity of a CAM data base is directly proportional to the number of tasks required by the system. In the ideal CAM system (Figure 10.1)

_____
* Real time refers to activities occurring in the present.

an extremely large and complex data base configuration is required. For an attainable CAM system (by present standards), the data base should contain those elements illustrated in Figure 10.3.

For the CAM system shown, only the engineering and manufacturing functions are under direct system control. Engineering design data enters the data base from computer-aided design programs used by various engineering departments. The design data and externally supplied manufacturing specifications are used by production/manufacturing engineering to develop NC part programs and other operational specifications. These are stored as NC production information in the data base. This engineering information flow is illustrated schematically in Figure 10.4.

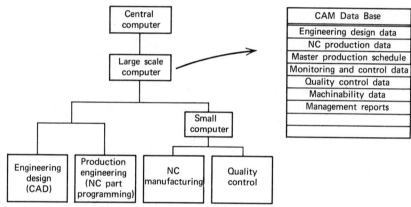

Figure 10.3   CAM data base for a computer based manufacturing system.

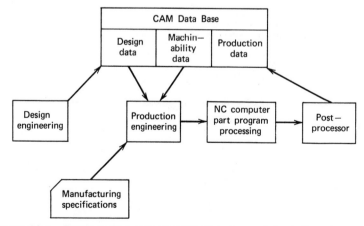

Figure 10.4   Engineering information flow to and from the data base.

10.3.2 Production Management

Once the finalized NC part program has been stored in the CAM data base, manufacturing priority dictates when the program is executed. The master production schedule, developed to reflect priority requirements of an individual company, is used by a scheduling program to control production. The scheduling program receives information concerning the status of production and makes the following information available (Reference 3):

1. Status of individual parts undergoing machining operations.
2. Status of each NC machine tool.
3. Actual production times versus scheduled production times.
4. Impending machine or system failures.

Based on this information, the scheduling program determines the *production load* for each operational machine tool so that established priorities are maintained.

When the scheduling program determines which NC part program is to be executed next, it places the data in a *ready bank* that is accessible, either directly or indirectly, to the numerical control machine's MCU. The program may also generate information concerning tooling, setup times, and manufacturing time for use by production personnel. This production management process is depicted schematically in Figure 10.5.

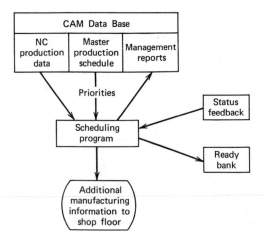

Figure 10.5 Production management information flow.

10.3.3 Manufacturing Control

The manufacturing control mode depends on the type of NC configuration used in the CAM system. Both DNC and CNC are investigated in detail in later sections. Consider the generalized manufacturing control system shown schematically in Figure 10.6. The manufacturing control program can be executed by a minicomputer at the NC machine site (CNC), a real time computer controlling a group of NC machines, or a large scale computer being linked to the NC machine tool via telecommunications lines (DNC).

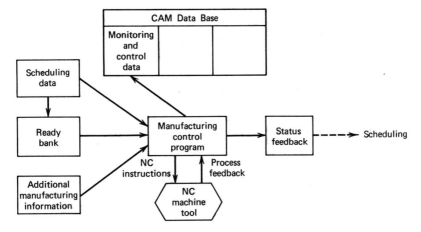

Figure 10.6 Manufacturing control information flow.

The manufacturing control program passes instructions to the NC machine based on data currently in the ready bank. NC data is obtained in blocks (subdivisions of the entire data structure or part program) from this ready bank as required. Individual NC blocks of information are passed to the machine. Process information and machine status are conveyed by the control program to the appropriate data evaluation modules.

The quality control function, illustrated in Figure 10.7, uses component specifications obtained from the CAM data base. The quality control program may receive manual input via interactive terminals at the inspection site or data obtained from NC inspection equipment. Regardless of the data source, the quality control program performs two distinct functions: it provides accurate validation of manufactured parts and initiates action by the manufacturing control system when tolerance exceptions occur.

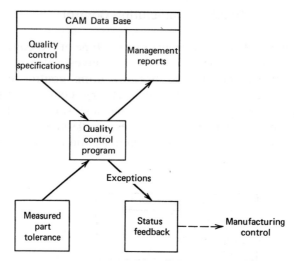

Figure 10.7   Quality control operations in CAM.

10.4 Computer Managed NC Systems

The manufacturing control system described makes use of a special NC machine that receives its instructions directly from a computer and communicates process information outside its own feedback loop. Because conventional NC equipment is incapable of performing these tasks, a modified NC system is required for CAM applications. Computer managed numerical control provides the two-way communication capability that is required in a computer based manufacturing system.

A computer managed NC system uses a digital computer to replace some or all of the logical functions performed by the conventional machine control unit. The computer enables the following benefits (Reference 4) to be realized:

1. The computer managed system receives its instructions directly from computer storage, thereby eliminating paper tapes and tape readers at the machine site.
2. Data concerning the ongoing machining process is fed back to the computer, where it can be stored or passed to a higher level management system.
3. NC logical functions are modularly designed using programmable software; hence, they are easily expandable and maintainable.
4. NC part programs, stored in the computer's memory, can be edited directly through an on-site console. The console also provides an excellent communication device between the computer and the NC machine operator.

5. The NC machine tool may be located many miles from the control computer. In some systems (DNC), the control computer manages a large number of NC devices.

Computer managed NC systems have been developed following two philosophies. Essentially, differences arise from the physical location, size, and function of the control computer. Two distinct systems have developed, both compatible with the CAM concept; however, at this time, no standard system exists, and even within a given system, a number of design differences are possible.

10.5 Computerized Numerical Control—CNC

Computerized numerical control (CNC) represents an important approach to computer managed NC in CAM. CNC, also referred to as *stored program numerical control* or *softwired numerical control*, has been defined by the Electronic Industries Association as:

> A numerical control system wherein a dedicated, stored program computer is used to perform some or all of the basic numerical control functions in accordance with control programs stored in the read-write memory of the computer.

A CNC system, therefore, replaces some or all of the hardware functions previously performed by the MCU with a *dedicated* computer, that is, a computer assigned to control a single NC machine.

10.5.1 Functional Elements of CNC

The difference between conventional NC equipment and CNC is the addition of a minicomputer as part of the machine tool controller, as shown in Figure 10.8. The controller of a typical CNC system contains two elements: pro-

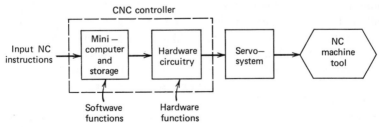

Figure 10.8 Computer numerical control (CNC) system schematic.

grammable or *softwired* computer modules that perform various control functions, and *hardwired* logical circuitry used for other control operations. The percentage of hardware functions as compared to software functions is extremely important to the overall performance of the system.

It is likely that in future CNC systems, all control functions will be efficiently performed using computer software. At the present time, however, software is generally not used for all functions, and a mild controversy exists in relation to the proper balance of hardware and software in CNC systems.

Two physical constraints limit the efficiency of a fully softwired NC control system. Analysis, interpolation, and logical software modules currently require relatively large amounts of computer storage (relative to that available in minicomputers). Additionally, current computer software cannot execute functions as rapidly as hardware circuits; however, neither problem is insurmountable, and some fully softwired controllers have been developed.

10.5.2 Software Interpolation

Interpolation in a CNC contouring system, like its counterpart in conventional NC equipment, is the most important computational task performed by the controller. Path interpolation is accomplished using *variable pulse rate circuitry* (VPRC) and *digital differential analyzers* (DDA); both are hardwired electronic circuits. To calculate the required interpolation path in a softwired system, these digital circuits must be simulated by a minicomputer program.

The problems involved with software simulation of hardwired circuits are best illustrated by considering the feedrate and path interpolation function required for three dimensional contouring, where three hardwired circuits (digital integrators) have to be simulated in the control software (Reference 5). Over twenty program instructions are required to perform the same task as a single circuit. The minicomputer would have to service each circuit every 12 microseconds on a time-shared basis to match the speed and resolution of a conventional NC controller. This task is not possible using the present generation of minicomputers.

To maintain comparable system response and resolution, the CNC designer has a number of possible alternatives. The interpolation function could be split between the minicomputer software and hardwired interpolators. The computer analysis program divides the tool path into segments, while the hardware circuits generate the points that keep the tool path within tolerance. Hence, software is used for coarse interpolation, and hardware (DDA) is maintained for fine interpolation.

A second method for software interpolation uses a *sampled data* technique (Reference 5). The computer becomes part of the position feedback control loop, interrogating the loop for position data at fixed time intervals, called the sampling frequency. During each sample, a computer program calculates the proper axis velocity from position data obtained from the feedback loop. Therefore, rather than performing calculations for each positioning pulse (pulse rates of 10,000/sec are not uncommon), the software takes samples 100 times/sec, a rate that is easily managed by the minicomputer.

Other interpolation techniques oriented toward high speed manipulation of binary arithmetic have also been developed. In general, the software systems discussed compare favorably with fully hardwired NC machine control units, although software interpolation does have weaknesses that can only be remedied through the development of more powerful computer technology.*

At the present time, a CNC system that is designed for both hardware and software would provide the best use of available resources. Less computer capacity is required for a mixed software/hardware system, allowing more room for storage of NC part programs.

10.5.3 Benefits of CNC

The most significant benefit of CNC is the flexibility of the system. Once a particular analysis function is provided by hardwired circuitry, it cannot be altered without physically rewiring the circuitry involved. A computer can be easily reprogrammed, and different programs can be efficiently stored for the execution of varied tasks. For this reason, a CNC system provides a multiplicity of options that could not be practically implemented in a conventional NC configuration.

Software computational features enable tool and fixture offsets to be computed and stored. Inch–metric conversions (see Metrification, Chapter Five) can be easily accomplished with the appropriate software routine. Absolute or incremental mode as well as point-to-point or contouring techniques can be selected. Sophisticated interpolation functions, such as cubic interpolation (Chapter Three), can be accommodated using software methods, and cutter compensation can be calculated as machining progresses. A large number of auxiliary functions can be programmed. Each of these features can be altered merely by rewriting the appropriate software routine. Hence, the incorporation of new features is greatly simplified.

---

* For example, the application of microprocessors to NC provides a new approach to the control function.

10.6 Direct Numerical Control—DNC

Direct numerical control (DNC) is a second major approach in computer managed NC systems. The Electronic Industries Association has defined DNC as:

> A system connecting a set of numerically controlled machines to a common memory for part program or machine program storage with provision for on-demand distribution of data to machines.

The important difference between DNC and CNC is the elimination of a *dedicated* computer by a larger computer that manages *many* machines on a time-shared basis.

A DNC system consists of a computer and four types of auxiliary equipment: bulk memory capable of storing different NC part programs for transmission to different NC machines; communication stations, including type address and cathode ray displays that provide an interface between the machine operator and the remote computer; telecommunications lines to transmit machine data to remote sites; and NC machines.

DNC encompasses the CAM philosophy more directly than does CNC. Each NC machine on the shop floor is controlled by a single centralized computer that receives feedback from the machines on a real time basis. Therefore, the computer can maintain an overview of all production operations under its direct control. Like CNC, direct numerical control is amenable to the hierarchical aspects of CAM. The DNC computer can be linked to a higher level computer that performs the production management and analysis functions associated with CAM.

10.6.1 DNC System Configuration

The general system configuration for direct numerical control is illustrated in Figure 10.9a. The central computer performs three tasks in relation to DNC; it retrieves part program instructions from bulk memory and disseminates the information to the appropriate machine. Then it controls the flow of information (in both directions) so that a request for NC instructions is immediately satisfied. This function is sometimes termed *traffic control*. Finally, the computer monitors and processes machine feedback for use in the CAM information loop.

Because it is essential that each NC machine receive its instructions at the precise time they are required,* the traffic control function can become

---

* Unlike other types of *time-shared* computer systems, DNC cannot tolerate protracted *wait times*. Even relatively small delays are detrimental to the response, accuracy, and efficiency of an NC machine.

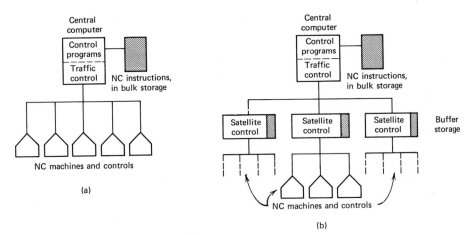

Figure 10.9 DNC system configurations. (a) Central computer control. (b) Satellite computer control.

difficult, particularly when the computer is located at a remote site, far from the NC equipment. The traffic control problem for large DNC systems can be alleviated by introducing a number of *satellite* minicomputers with their own storage buffer. Each satellite computer receives large blocks of NC data from the central system and stores this information in its buffer. The task of disseminating specific data at the proper time is left to the minicomputer. Hence, the burden of traffic control is divided between the central computer and each of its satellites. A schematic for such a system is shown in Figure 10.9b.

Although the DNC computer(s) represents the highest level(s) of control, each NC machine may have a control unit on site to interpret the machine commands and act upon them. Two methods, discussed in the following sections, are available to accomplish this task.

10.6.2 Behind-the-Tape Reader Systems

The tape reader is an integral part of conventional NC equipment. All machine commands, coded on perforated tape, are passed to the MCU through the tape reader. In a behind-the-tape reader (BTR) system, the only change made to the NC machine is the replacement of the tape reader by a telecommunication line connected to the DNC computer.

As illustrated in Figure 10.10, the BTR system has all the attributes of the DNC systems discussed in the previous section. The traffic control dispatches individual blocks of NC machine instructions over telecommunication lines to

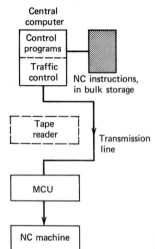

Figure 10.10   A Behind-the-tape reader (BTR) system schematic.

the MCU. The control unit has two buffer registers, one used to receive data from the computer and the other used during instruction processing. Since this arrangement is also typical of conventional machine controllers, little, if any, internal modification need be made to the MCU.

Consider a typical BTR system operating sequence,* using a configuration similar to the one shown in Figure 10.9a. A portion of the central computer's core memory is set aside for double 256 character buffers (512 bytes) for each machine tool. The NC instructions for a given machine tool are transferred in blocks of 256 characters from bulk storage to the core buffers. This ensures that a minimum of 256 characters of data will be available to the machine tool.

In the system under consideration, the control program consists of two traffic control routines—the *machine scanner* and the *auxiliary scanner*. The machine scanner manages the flow of information between the core memory buffers and the NC equipment. The auxiliary scanner processes a two-way flow of control information concerning the operation of the machine tool (Reference 5).

When the MCU transmits a request for NC instructions, a signal is sensed by the machine scanner and one character of an NC block is sent to the machine. The scanner then proceeds to poll the rest of the machine tools, sending one character of data to each device that has posted a request. Upon returning to the first NC machine, the scanner will transmit the next character of data, if a request for data is still posted. This process is performed in a continuous

---

* The system described is based on a General Electric CAM system (References 2 and 5).

manner so that each machine is serviced in order. Since it takes approximately 5 μsec to transmit one character of data, data can be transmitted to fifteen individual control units at a rate of 1250 characters/sec. As characters are transmitted, one of the double core buffers is emptied. Transmission continues from the second buffer, while the original is refilled.

At intervals during the machine scanning cycle, the auxiliary scanner is used to process messages, such as block numbers and other information not directly associated with NC machine commands. The use of two scanning routines provides an efficient method of traffic control.

An important aspect to note is that control routines in a BTR system merely transmit standard machine instructions. The hardwired features of the conventional MCU are not replaced by software; that is, interpolation and logical functions are performed by the machine tool controller.

10.6.3 DNC Systems with Special Machine Control

Many DNC systems have been designed so that the computer (theoretically, a satellite minicomputer) replaces some or all of the functions normally performed by the machine control unit. Like the CNC systems discussed previously, DNC software interpolation can be executed entirely by computer, or the interpolation function can be split between computer software and MCU hardware.

Figure 10.11 illustrates two basic configurations for DNC systems which introduce software into the control process. In Figure 10.11a, digital data is transmitted from the central computer to an interpolating buffer unit* at a rate of approximately 160,000 characters/sec. The buffer unit uses interpolation logic to establish axis feedrates and generate the proper number of pulses for axis motion. A separate interpolating buffer unit is used for each NC machine tool under the control of the DNC system (References 6 and 7).

The DNC system shown in Figure 10.11b uses a coarse interpolator that calculates the segments of cutter path motion required for each NC machine tool. The interpolator receives data from the computer and calculates one coarse segment of motion for each machine tool on a sequential time-shared basis. Fine interpolation is performed by a special machine control unit at the machine site. In a typical system of this kind, the time-shared coarse interpolation technique can accommodate feedrates as high as 10 mpm on all NC machines.

---

* The term *interpolating buffer unit* is generally associated with DNC systems developed by the Sundstrand Machine Tool Company.

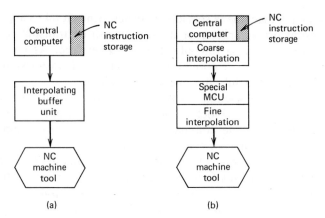

(a)   (b)

Figure 10.11   DNC systems with special machine control.

10.6.4 A Comparison of DNC Systems

Any comparison of BTR and special machine control DNC systems should consider the relative merits of five factors: *data management capability*; *reliability*; *flexibility*; *complexity*, and *economic criteria.*

The data management capability of a DNC system encompasses system input and data format, data entry procedures, and the ability to store information for subsequent management reporting. Both BTR and special machine control systems make use of a remote computer. The major difference between the systems is that the BTR uses the existing MCU. Therefore, BTR input is in character format and no management data storage capability exists. The special machine control system uses binary input and provides for manual data entry and management data storage.

As in other complex electromechanical systems, DNC reliability is measured by the least reliable element. If control system failure is considered as a criteria, a BTR system is as reliable as the MCU itself. Special machine control systems make use of digital computer software that generally has high reliability (Reference 5).

A conventional MCU is inflexible because of the hardwired logical functions it performs. For this reason, BTR systems have a relatively rigid configuration. Special machine control systems, however, with a high proportion of softwired control, can be easily reconfigured to accommodate new features. Because flexibility is inversely proportional to complexity, the BTR system uses much simpler hardware and software than the special machine control system.

From an economic standpoint, two factors must be considered: the cost of the DNC system and its ability to increase profits. The cost of special machine control systems is high due to expensive hardware and software systems, but its potential for profitability is also high. BTR systems are relatively inexpensive since existing NC devices are used with minimal modification. However, because overall control is limited to the replacement of the tape reader, long range profitability will not be as high.

In summary, BTR is a system that enables the implementation of DNC with relatively little modification of the user's equipment. It provides the first links to CAM but lacks certain important features. Special machine control systems make use of the most sophisticated DNC technology, and provide a more sophisticated link with an overall computer based manufacturing system. The choice of the "right" DNC system depends upon the present needs and future goals of the user.

10.7 Programmable Machine Error Compensation

The computer has made possible capabilities for error compensation and prevention in computer managed NC systems that may someday overshadow sophisticated control functions. In this section three important computer analysis features related to error compensation and prevention are considered.

10.7.1 Axis Calibration

The accuracy of a conventional NC machine (Chapter Five) is directly related to the accuracy of each element in the system. A leadscrew that is designed with a tolerance of 0.02 mm cannot be expected to position to an accuracy of 0.01 mm in conventional NC equipment. However, a computer managed NC system can enable a machine tool to produce parts with accuracies and tolerances that are better than those of the machine tool itself!

In Chapter Five we discussed many factors contributing to positioning error in an NC machine tool. Lost motion due to backlash, windup, deflection, and frictional effects all influence machine accuracy. In addition, temperature variance, actuation system inaccuracy, component wear, and overall control system eccentricities contribute to the accuracy problem. For conventional NC devices, the only solution to accuracy problems is to design each component within close tolerances in an effort to minimize all factors. To keep the machine within specifications during use, periodic adjustments must be made to compensate for component wear and drift. Such adjustments can result in a large amount of down time for the NC machine.

Computer managed NC systems offer an alternative approach. By combining the features of an extremely accurate measurement device, such as a laser interferometer and digital computer, it is possible to calibrate each axis of the NC machine. *Axis calibration* consists of the following steps:

1. A laser interferometer is mounted on the machine slides for each axis of motion so that the exact position of the axis can be determined.
2. Using standard NC instructions, each axis is positioned at an arbitrary number of points across the entire range of axis travel.
3. The deviation between the command position (i.e., the position requested by the NC instruction) and the actual position measured by the laser interferometer is recorded by a computer.

This sequence may be repeated for different levels of force loading, resulting in a record of error versus command axis position (Figure 10.12*a*) that is available for each axis.

Since the resultant machine inaccuracy is stored in the computer's memory, each command signal to the servo system can be automatically modified so that the actual command position is achieved. Referring to Figure 10.12*a*, a positioning command to the absolute coordinate 30.000 will result in an error of 0.006 mm. When a command to position at 30.000 is received, the computer *looks up* the corresponding error and compensates the command to 29.994 mm.

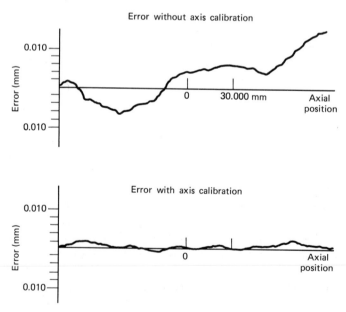

Figure 10.12   Axis calibration and error compensation for computer managed NC.

Over the entire axis range, with axis calibration the error is greatly reduced, as shown in Figure 10.12*b* (Reference 8).

The axis calibration process can be performed at given time intervals during the life of the NC machine tool. A new error curve is stored by the computer and changes in accuracy (for whatever reason) can be accommodated. Computer compensation for known system errors also enables the NC machine manufacturer to relax extremely high tolerance requirements, resulting in reduced cost without any decrease in system accuracy.

10.7.2 Real Time Error Compensation

If a laser interferometer were permanently mounted along each axis of motion, exact position data, independent of standard transducers in the servo system, could be fed back to the computer software. Positioning error would be noted by the computer, and compensatory action initiated. Effectively, the NC machine axes would be calibrated on a continuous basis. Precision measuring devices can also be used to measure the pitch and yaw of the work-holding table.

10.7.3 Maintenance Capabilities

A computer which manages the NC machining process can also be used in the location and diagnosis of other system problems. The axis calibration technique provides the computer control system with an ongoing record of positioning errors that can be evaluated to determine whether maintenance is required. Even preventative maintenance, a task that is crucial to every NC machine, can be enhanced using the computer and NC machine feedback.

Based on predefined maintenance requirements, a CAM system supplied with maintenance data can produce periodic (e.g., weekly, monthly) requests for maintenance, listing regular preventative maintenance chores as well as special maintenance. For example, repeated adjustment of a particular axis position (using axis calibration) can indicate a problem in actuation system components. As part of a maintenance request printout, a notation can be made concerning the problem, and a listing of possible error source locations can be provided (Reference 9).

Some computer managed NC systems have been programmed to make real time maintenance checks during production runs. If the computer detects any performance deterioration, a variance report is issued, and the NC machine need not be taken out of production for minor maintenance tests. In the event of impending catastrophic failure, the control software is programmed to shut down the system, averting costly damage to the machine and/or the workpiece.

10.8 Monitoring in a CAM System

The introductory section of this chapter outlined an ideal hierarchical structure for a computer based manufacturing system. Referring again to Figure 10.1, the inputs to the overall CAM system fall into two broad categories: numeric data relating to management and analysis functions, and machine or operation data required directly from an ongoing process. Information in the first category is supplied directly by personnel; it is generally not required that data be input in real time or that data be transmitted continuously. Information in the second category is generally a real time, continuous stream of data that is automatically obtained using sensing devices of various levels of complexity. It is the second category of data acquisition in a CAM system that is termed *machine monitoring*.

Machine monitoring has been defined as: "the computerized acquisition of machine, process, product, and/or operator status information from sensing elements attached directly to a machine" (Reference 10). In relation to the computer managed numerical control systems discussed earlier, machine monitoring includes information flow in both directions; that is, technical data is sent to an NC machine and compared with monitored data fed back from the machine. However, the concept of machine monitoring may also be applied to plant facilities management, pollution control systems, production and inventory control, as well as to a broad range of specialized communication functions.

10.8.1 Monitoring Requirements

The most important element in the machine monitoring process is a computer capable of isolating signals from the machine and transforming them into meaningful information. Hence, the monitoring computer must be interfaced with one or more sensory devices in a real time mode. In Chapter Four, we discussed a number of electronic and electromechanical transducers used in NC systems. Any device that has a voltage or current output can be sensed by a computer.

Because machine monitoring is not strictly limited to NC applications, a wide variety of data types (e.g., piece counts, out of tolerance conditions, machine status) can be monitored by the computer. The quantity of data and speed with which it is presented to the computer dictate the level of performance required of the monitoring system.

Low performance monitoring requirements are found in applications in which relatively little change in condition is encountered. Generally, a particular condition changes state at a rate of much less than once a second. An example of a low performance digital sensing system might be a photo diode counter which indicates components traveling along a slow moving conveyor.

Medium performance monitoring requirements include applications in which the rate and peak values (with respect to time) of various process parameters are required. State changes occur at a rate of 1 to 60 times/sec. High performance monitoring considers events that occur at intervals of less than 1/60 sec. In such systems the digital address or analog profile of a transient phenomenon can be determined. Consider, for example, a measurement system that provides machine slide position with respect to time at 1/10,000 sec intervals. The resultant data yields a near continuous profile of slide position as well as velocity and acceleration.

10.8.2 Techniques for Signal Acquisition

The complexity of an NC machine tool–computer communication interface depends upon the number and type of feedback processed, and the number of machines that the computer must monitor. Although feedback may originate at sensing devices that produce digital, analog, or mixed (digital and analog) signals, we have seen that the computer evaluates data in digital form. Therefore, all feedback must be converted to pulse type (digital) signals (Figure 10.13a).

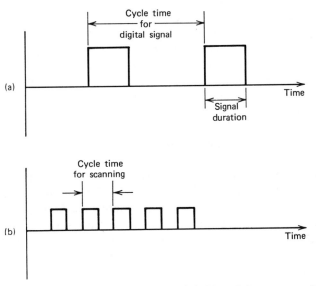

Figure 10.13 Scanning for digital signals. (a) Signal frequency. (b) Scanning frequency.

In a typical computer managed NC system, the control computer may have responsibility for one or more NC machines. Each machine generates digital feedback that must be recognized and then evaluated by the control computer. The computer then *scans* the input from each machine at a rate that insures that no data will be missed. For the signal shown in Figure 10.13*a* a recommended *scanning time* would be approximately 2.5 times as fast as the minimum signal duration. The scanning time, indicated in Figure 10.13*b*, is the increment of time during which the computer is *looking* at a particular input from a given machine.

The monitoring techniques used in actual DNC and CNC cluster systems are beyond the scope of this text. Computer managed NC systems generally make use of techniques originally developed for time-sharing systems for data processing computers.

10.9 Trends in Computer-Aided Manufacturing

Computer-aided manufacturing is currently a relatively unstructured branch of technology. Although prototype manufacturing systems have already been developed, it will take years before the diverse disciplines that contribute to all elements of CAM are coordinated and produce a fully operable system. The spread of computer based automation into a large segment of industry will undoubtedly occur, but the process will take decades to accomplish.

The most significant achievements in CAM have been the interface of numerical control with a hierarchical computer system for machine control, process monitoring, and NC part program development. The engineering design function (CAD) has been successfully coupled with NC oriented CAM systems, providing a highly automated engineering production system. DNC and CNC systems will incorporate a greater degree of adaptive control, resulting in computer managed AC systems with programmable optimization capabilities.

It is predicted that computer managed NC systems will be expanded to include industrial robots, automated warehousing systems (Reference 1), and NC machine tool clusters connected by conveying systems for expeditious material movement. Industrial robots, capable of computer coordinated visual and tactile senses, will provide a flexible interface between the conveying system and the NC work station. Automated warehousing facilities will be used to satisfy computer generated requests for raw materials and, at the other end of the production process, finished products for shipping.

References

1. Anderson, R. H., "Programmable Automation: The Bright Future of Computers in Manufacturing," *Datamation*, vol. 18, December 1972, pp. 46–52.
2. Harrington, J., *Computer Integrated Manufacturing*, The Industrial Press, New York, 1973.
3. "Computer Managed Numerical Control," *N/C Handbook*, 3rd ed., L. J. Thomas, Ed., Bendix Corporation, Industrial Controls Division, 1971, pp. 203–10.
4. Budzilovich, P. N., "Computerized NC—A Step toward the Automated Factory," *Control Engineering*, vol. 16, July 1969, pp. 62–8.
5. Mesniaeff, P. G., "The Technical Ins and Outs of Computerized Numerical Control," *Control Engineering*, vol. 18, March 1971, pp. 65–84.
6. Caruthers, P., "A New Concept in DNC," *The Proceedings of the Tenth Annual Meeting and Technical Conference*, Numerical Control Society, New York, 1973.
7. *IBM System/370 Machining and Display Application Program (MDAP) for IBM 5275 Direct Numerical Control Station, General Information Manual*, IBM Corporation, publication GH20-1293, April 1973.
8. "New Potentials in Computerized NC," *N/C Application Guide*, Allen-Bradley Company.
9. Cope, K. L., "NC Maintenance for the Future," *American Machinist*, vol. 117, December 10, 1973, pp. 71–3.
10. *Computer Aided Manufacturing/Concepts in Machine Monitoring*, IBM Corporation, publication GE-20-0410, June 1973.

Index